STEM, Social Mobility and Equality

"This book provides compelling evidence not only of the failure of UK social mobility policy but why it fails. This book is essential reading for all those concerned about inequalities in education. It combines a carefully considered genealogical analysis of the social mobility of university students studying STEM subjects with the powerful message that social mobility policies have been woefully inadequate."

—Diane Reay, *University of Cambridge, UK*

"For a generation, the ideas of social mobility and science have intertwined in British governments' education policies. This book combines big ideas and minute qualitative research to question a good deal of what we—and governments—have taken for granted. It's a book which should be read by every education policy maker who wants to think seriously about education, technology and class."

—Chris Husbands, Vice Chancellor, *Sheffield Hallam University, UK*

"A carefully executed study that critically examines the realities of STEM for the possibility of achieving social mobility and equality. This book provides rich insights into the identities of chemistry undergraduates and the experiences of staff with Athena Swan and the challenges of equity policies. A quality and timely sociological contribution to science education research and scholarship."

—Billy Wong, *University of Reading, UK*

Kate Hoskins • Bernard Barker

STEM, Social Mobility and Equality

Avenues for Widening Access

palgrave
macmillan

Kate Hoskins
Education
Brunel University London
Uxbridge, UK

Bernard Barker
Education Leadership & Management
University of Leicester
Leicester, UK

ISBN 978-3-030-49215-1 ISBN 978-3-030-49216-8 (eBook)
https://doi.org/10.1007/978-3-030-49216-8

This Palgrave Pivot imprint is published by the registered company Springer Nature Switzerland AG.
The registered company address is: Gewerbestrasse 11, 6330 Cham, Switzerland

KH: For Ian, the most amazing, kind and supportive brother I could ask for. Thank you.

BB: For Margaret and David Killingray, lifelong friends and supporters

Acknowledgements

Our warm thanks are due to the staff and students at Marsden University (pseudonym) who encouraged and facilitated the research, especially those who agreed to participate in the interviews and the students who further assisted us by supplying vital genealogical data about themselves and their families. The result is an unusual portrait of undergraduate chemists and their families.

Huge gratitude to Bruce Gilbert for providing Bernard with a lifetime of discussion about science, science education and the evolving character of higher education. We are indebted to him for many insights and great friendship. Great thanks also to Lynn Mayes and her extraordinary colleagues at Queen Katharine Academy in Peterborough. They have done more to improve social mobility for their students than anyone else we know.

We also warmly acknowledge the stimulus we have gained from Diane Reay's formative contribution to the debate about social mobility and class; and from Phillip Brown's recognition that too much policy depends on ignoring 'sociology's inconvenient truth'.

Our families have been silent partners in this testing but rewarding project and we offer our love and thanks for their patience, endurance and shrewd remarks. Damien, Zachary and Dylan; Ann, Irena, Carmen, Conrad and Marcel—you've been great.

Contents

1 Introduction 1

2 The Contemporary Social Mobility and Equality Policy Context: Framing the Problem 19

3 Making Chemists 61

4 Equality Policies and Initiatives at Marsden 89

5 The Limits of Equality Policy 113

6 Conclusion 133

Appendix 147

References 151

Index 169

About the Authors

Kate Hoskins is a Reader in Education at Brunel University. Her academic publications are concerned with issues of policy, identity and inequalities in further and higher education. In her book Women and Success: Professors in the UK academy (Trentham Books, 2012) she used life history interviews with 20 female professors to reveal the persisting inequality facing senior women working in higher education. In her co-authored book *Education and Social Mobility: Dreams of Success* (IOE, Trentham Books, 2014) she and Professor Barker draw on data collected through paired interviews with 88 15–18 year olds in two secondary schools to explore their aspirations for the future. In 2017 she published a book entitled *Youth Identities, Education and Employment: Exploring Post-16 and Post-18 Opportunities, Access and Policy* (Palgrave Macmillan, 2017), which is based on in-depth qualitative interviews and focus groups with teachers and students in a case study school and investigates how policy, family background, social class, gender and ethnicity influence young people's post-16 and post-18 education access.

Bernard Barker is Emeritus Professor of Educational Leadership and Management at the School of Education, University of Leicester and Chair of Governors at Queen Katharine Academy in Peterborough. He was educated at Eltham Green School in south London and Gonville and Caius College, Cambridge. He progressed to York University where he researched the Labour movement in the years immediately before and after the First World War. Between 1971 and 1999 he taught at schools in

Hertfordshire, Cambridgeshire and Leicester, including 17 years as the principal of Stanground College, a large community comprehensive in Peterborough. His book *Rescuing the Comprehensive Experience* (Open University Press, 1986) reaffirmed the importance of comprehensive education in the threatening environment of the Thatcher years. His recent publications include *The Pendulum Swings: Transforming School Reform* (Trentham, 2010), *Human Resource Management in Education: Contexts, Themes and Impact*, with Justine Mercer and Richard Bird (Routledge, 2010), *Education and Social Mobility: Dreams of Success* (IOE, Trentham Books, 2014) with Kate Hoskins and *Busking Latin: A Memoir* (The Stamford Press, 2019).

Abbreviations

A level	Advanced Level
ARM	Advanced Resource Managers
ASC	Athena Swan Charter
ASE	Association for Science Education
BAE	British Aerospace Systems
BAME	Black, Asian and Minority Ethnic
BBC	British Broadcasting Corporation
BCS	British Cohort Study
BERA	British Educational Research Association
BHPS	British Household Panel Survey
BITC	Business in the Community
BME	Black and Minority Ethnic
BSA	British Sociological Association
CUPESSE	Cultural Pathways to Economic Self-Sufficiency and Entrepreneurship
CV	Curriculum Vitae
DCSF	Department for Children, Schools and Families
DfE	Department for Education
DFES	Department for Education and Skills
ECG	Electrocardiogram
ECU	Equality Challenge Unit
EHRC	Equality and Human Rights Commission
ESRC	Economic and Social Research Council
EU	European Union
FSM	Free School Meals
GCE	General Certificate of Education
GCSE	General Certificate of Secondary Education

GIST	Girls into Science and Technology
HE	Higher Education
HEI	Higher Education Institutions
HESA	Higher Education Statistical Agency
HMG	Her Majesty's Government
HMT	Her Majesty's Treasury
HP	Hewlett Packard
HSCC	Health and Social Care Committee
IT	Information Technology
KCL	King's College London
LGBO	Lesbian, Gay, Bisexual Outreach
LGBT	Lesbian, Gay, Bisexual, Transgender
LGBTQI+	Lesbian, Gay, Bisexual, Transgender, Questioning (or queer), Intersex
M. Chem	Master of Chemistry
MCS	Millennium Cohort Study
MP	Member of Parliament
NCDS	National Child Development Study
NIH	National Institutes of Health
NSF	National Science Foundation
NUS	National Union of Students
OECD	Organisation for Economic Cooperation and Development
PFAP	Panel on Fair Access to the Professions
PhD	Doctor of Philosophy
POLAR	Participation of Local Areas
REC	Race Equality Charter
REF	Research Excellence Framework
ROSLA	Raising of the School Leaving Age
SEB	Socio-Economic Background
SEND	Special Educational Needs and Disability
SET	Science, Engineering and Technology
SMC	Social Mobility Commission
SMCPC	Social Mobility and Child Poverty Commission
STEM	Science, Technology, Engineering and Mathematics
STEMM	Science, Technology, Engineering, Medicine and Mathematics
TISME	Targeted Initiative on Science and Mathematics Education
UCAS	Universities and Colleges Admissions Service
UK	United Kingdom
US	United States
WISE	Women into Science and Engineering

CHAPTER 1

Introduction

Abstract In this chapter, we contextualise our project and relate the case study to the wider educational policy concern in England, which emphasises equal opportunity policies that aim to improve social mobility for all (DfE, *Unlocking Talent, Fulfilling Potential: A Plan for Improving Social Mobility Through Education*, CM 9541, 2017), and Athena Swan's (2005) concern with gender equality, fairness and the workforce's welfare. We provide a rationale to justify our choice of case study within the STEM (Science, Technology, Engineering, Mathematics) subjects, which we argue is because of the importance given to these disciplines by policymakers (HMG 2012: House of Lords, 2012). We provide an in-depth discussion of the methodology and methods informing our study and justify taking a case study approach. In doing this, we point out the wider appeal of our case study findings to fields beyond chemistry and show how our work provides insight into studying STEM subjects more broadly.

Keywords STEM • Social mobility • Equality opportunities • Policy

A casual conversation with a former Pro Vice-Chancellor of Marsden University (pseudonym) gave an unexpected introduction to a top-rated Russell Group chemistry department, where over 90 per cent of the research is rated as 'internationally excellent' (Research Excellence

K. Hoskins, B. Barker, *STEM, Social Mobility and Equality*,
https://doi.org/10.1007/978-3-030-49216-8_1

Framework [REF] online). The department believes passionately in social inclusion and equal opportunities for students and staff. It aims to improve and enhance every aspect of its work so that everyone is empowered to pursue their studies and careers without hindrance from background variables, including social class and disadvantaged circumstances, in addition to the characteristics protected by the Equality Act (2010), namely age, disability, gender reassignment, pregnancy and maternity, race, religion and belief, and sex and sexual orientation. It has engaged on a sustained basis with the Athena Swan Charter (ASC). The Charter assists institutions to build a culture that values diversity and equality; and encourages cultural and systemic changes that make a real impact on the lives of staff and students (Bhopal 2018; AdvanceHE 2019).

Our chance contact presented, therefore, an unusual opportunity to examine troubling concerns arising from government policies to promote social mobility, *Science, Technology, Engineering and Mathematics* (STEM) and equality more generally. There is a remarkable gulf between ambitious policy plans in this arena and the limited outcomes achieved, especially in science-related disciplines and occupations. Initiatives and legislation intended to ensure equal opportunities, regardless of an individual's background characteristics, have achieved less than many hoped. This is a case study in higher education (HE), therefore, that traces the STEM student trajectory through family, school and university (based on interviews with undergraduates); and examines the adoption, implementation and relative success of the ASC in the Marsden chemistry department (based on interviews with staff). Departmental plans and submissions for charter recognition are examined and discussed in the context of staff perceptions of the Athena Swan process.

The overall aim of our book is to better understand how students' identities and aspirations are formed and developed; and in light of that understanding to assess the effectiveness of strategies to enable career progression and mobility; to extend and enhance equal opportunities; and also, to identify improvements that may help institutions become more fully inclusive. This case study has wider significance, although the thick descriptions and detailed knowledge we have assembled here do not lend themselves to generalization. The aim is not to generalize, but to identify and describe influences and relationships and to formulate hypotheses about mobility processes that can be compared with the results of similar enquiries (Bertaux 1995). The Marsden findings are relevant to the wider debate because they identify areas of inclusion that work well and draw critical attention to systems and processes that need improvement.

The Problem

Since at least *Labour's Plan for Science* (Wilson 1963), human capital theory (Goldin 2014) has justified an almost continuous effort to educate young people for an advanced, science-based economy. The aim has been to make the country richer by opening routes to new, rewarding careers, with upward social mobility for highly skilled individuals who choose to work in STEM subjects. The widespread perception that social mobility is open to everyone who aspires to success, works hard and seizes opportunities, has further encouraged multiple attempts to raise aspirations, improve STEM skills and create a meritocratic society (Hoskins and Barker 2019).

These goals have proved elusive, however, despite reforms that were supposed to enable disadvantaged children to overcome their backgrounds (Department for Education [DfE], 2010, 2017), and the rapid expansion of HE, designed to widen participation, especially for under-represented groups. Equal opportunities, apparently secured by the Equality Act (2010), seem less open to all when closely examined. There is evidence, for example, that disadvantaged students perform consistently less well at school than their peers (Cook 2012; Andrews et al. 2017) and that top professions are dominated by people with private school backgrounds (Social Mobility Commission [SMC] 2016; Friedman and Laurison 2020). Women are 'vastly under-represented among top earners' (Fortin et al. 2017: 1) and a substantial pay gap persists across all occupations (Olsen et al. 2014). Race and ethnicity are equally powerful in shaping people's opportunities (Powell 2019) and what individuals believe is possible and desirable (Archer et al. 2012).

The drive to promote STEM subjects at school and university seems to have been equally unsuccessful. As the President of the Royal Historical Society says,

> Especially from the late 1950s, when economists and industrialists began to bear down on educational policy, the public discourse about subject choice has been dominated by the alleged insufficiency of science (or what we now call STEM) and policy has been aimed almost exclusively at remedying that insufficiency. How misleading this discourse is as a guide to actual behaviours is clear from the brute fact that there has been a nearly continuous

> swing away from science from the early to mid-1960s up until the last few years. (Mandler 2017)

The expansion of HE seems to have made little difference to relative life chances, while participation in STEM courses has been stable in terms of undergraduate backgrounds and ethnic group (Smith 2011). Young people, regardless of their origins, seem resistant to attempts to steer their choices and decisions; and appear driven by background influences that do not appear on the policy-makers' mobility map.

The Case Study

At South Park and Felix Holt, two high-performing academies, we chose a best-case scenario for an earlier social mobility study, designed to test the government's assertion that improved educational access, combined with high-quality teaching, will enable young people to transcend disadvantaged backgrounds and achieve hard-earned success. Interviews with 16 and 18-year olds led to the conclusion that government expectations are unduly optimistic (Hoskins and Barker 2014). We found little evidence that school reforms created favourable conditions for upward mobility; and our data confirmed the importance of family, community and other deep structural influences in shaping aspirations and goals.

Individual and family employment trajectories emerged as intimately related, with social, economic and cultural capital transmitted through the three generations (grandparents, parents, young person) investigated (Hoskins and Barker 2019). Interests and occupational dispositions were linked, with participants' progress making sense in the context of their extended family history and historic changes in the local and national job markets. These findings suggest the government's reliance on an over-simple, individualist approach to education and vocational choice has contributed to the sustained failure of initiatives to increase social mobility in general, and to expand the supply of STEM graduates in particular. Policy-makers are trapped in their own assumptions, especially with the idea that family background is a handicap to be overcome, not a shaping influence on young people's education and vocational choices.

This investigation begins therefore from a different perspective, grounded in our earlier work, and uses a genealogical method to examine the family and policy influences on undergraduate chemists. How have equal opportunity policies (e.g. widening participation; the ASC) combined with family networks, teachers, schools and universities to influence the attitudes and aspirations that condition academic and vocational choices and pathways in a STEM subject? What barriers have the students encountered so far, even in the positive environment of a department strong in Athena Swan values, and how might they be overcome? The influence of gender, ethnicity and social class are considered in relation to their role in enhancing or limiting aspirations to study chemistry.

A genealogical, qualitative case study, as recommended by Bertaux and Thompson (1997), is a promising way to examine the influence of family networks, habitus and dispositions on the academic and vocational pathways selected and followed by undergraduate chemists at a Russell Group university. The approach opens territory not easily accessed with the statistical methods familiar from so many social mobility studies. Labour market economists, for example, have prioritised the use of large-scale quantitative surveys to chart income differences between fathers and sons in successive birth cohorts, and so to measure intergenerational fluidity. But although such surveys emphasise the relative importance of individual male incomes in social change, they often neglect other dimensions, especially advances in women's employment status and earnings, and the role of family networks in holding and transmitting social, economic and cultural capital through the generations (Lambert et al. 2007; Bertaux and Thompson 1997). Quantitative surveys and data may show broad movements and trends, like the fluctuating lines on an electrocardiogram (ECG) print out, but they provide few insights into the decisions made by individual agents and the underlying social processes that influence choice and action.

A genealogical approach has distinctive advantages when the aim is to understand how young people's aspirations and identities are formed, and to investigate how their achievements relate to family and science capitals as well as to their social class background. We should acknowledge that 'mobility is as much a matter of family praxis as individual agency,' with families rather than individuals carrying social and occupational status (Bertaux and Thompson 1997: 7). This recognition broadens and deepens our understanding of mobility processes. The unit of sociological investigation becomes a section of an unlimited texture of kinship ties, rather than an individual or even the nuclear family (Bertaux 1995). In this context, occupations cease to be the sole indicator of social position and interest shifts from fathers and sons to the wider family network over time. Housing, education and culture are also taken into account. These are considered in light of a multi-generational perspective that shows how some families are able to build on advantages they already have to move up in successive generations (Mare 2011). The role of women in raising children and transmitting family influences is clarified while the family itself is recognised as the principal channel for the transmission of language, culture, status and values (Thompson 1997).

Genealogical methods have been developed less than they might have been, however, because most mobility studies have relied on data for just two generations of family members. This is because information about grandparents and great grandparents can be difficult to collect and may not be readily available (Chan and Boliver 2013; Erola and Moisio 2007). The two-generational model seems to have prompted an over-estimate of intergenerational mobility and has encouraged an unhelpful emphasis on fathers and sons (Pfeffer 2014). Our understanding of family networks is considerably enriched when three generations are included and the multi-generational effects of processes unfolding within the nuclear family can be examined (Mare 2011). There is a growing understanding of the role of grandparents. Analysis of data from three British cohort studies has shown, for example, that absolute mobility rates vary substantially by grandparents' social class and that even when parents' class is taken into account, there is a statistically significant association between grandparents' and grandchildren's class positions. The study concludes that the grandparent effect on social mobility is considerable and operates throughout the class hierarchy (Chan and Boliver 2013).

A set of individual, semi-structured life history interviews, each an hour in duration, is at the heart of this enquiry. Individual undergraduate chemists present perceptions and interpretations of their family background; their experience of primary and secondary school as well as their choices of course and university; and the shaping of their current aspirations and expectations for the future. The intention was to gather trustworthy information about each student and six other members of her or his family by adopting a neutral stance in clarifying and accepting the accounts offered. Where memories were vague or incomplete, especially about the grandparents, participants' recollections were recorded as they were given. This is a highly efficient means of data collection about a kinship network and produced thick descriptions rich in family history although there are costs in validity and reliability (Bertaux 1995). It would have been difficult within the resources of our research to have triangulated participants' stories with other family members but there is the compensating thought that the personal accounts gained strength from the interpretations and descriptions offered.

Sample & Access

The study is based on interviews with a sample of 11 undergraduate students at Marsden University pursuing 3- or 4-year degree courses in chemistry; and with 8 staff members of the chemistry department. Student participants are self-selected volunteers who responded positively to a general invitation to take part in the research. The sample is not representative of undergraduate chemistry students at this or any other university but nevertheless provided ethnologically dense knowledge and 'thick' descriptions that can lead to interpretive ideas, hypotheses and concepts (Geertz 1973). Staff participants were nominated by the head of department and represent the current outlook in chemistry.

We explained the aim and objectives of the project and outlined the time commitment required. The eventual opportunity sample of students included variations in gender and socio-economic background. The gender composition of the sample was set by the individuals volunteering but Marsden chemistry cohorts are near parity in terms of male/female representation. We hoped to achieve diversity in relation to ethnicity, but this was limited as course participants tended to be white British or white European students. This lack of ethnic diversity is a limitation for the study and is discussed in Chap. 4 in relation to the policies and initiatives implemented to enhance minority ethnic representation in the department.

The student sample is smaller than originally planned but has yielded rich data consistent with other, much larger mobility studies, and provides insight into personal and background influences on academic and career decisions. These included questions about the background, occupational trajectory, dispositions and influence of parents and grandparents. Participants also completed a social mobility questionnaire about their parents' and grandparents' education, work history and perceived social class background. They were asked in addition to match family members with job/status descriptors on the Registrar General's Occupational Classification Chart. The data collected includes rich material on the lives and mutual influence of 77 individual family members, including the participants themselves. The findings based on the undergraduate interviews, as reported in Chap. 3, help further illuminate the operation of Athena Swan in the chemistry department (as detailed in Chap. 4). Student details and background are summarised in Appendix 1: Chemistry Students: Family Occupations and Class Identities.

The staff sample comprises 8 individual interviews, each of approximately 40 minutes. These capture staff perceptions of current progress towards improving equal opportunities and reducing obstacles for the undergraduates, postdoctoral students and early career researchers who seek to achieve educational success and career mobility in chemistry.

Ethics

The study complies with the ethical protocols set out by the British Education Research Association (BERA) (2018) revised ethical guidelines and the British Sociological Association (BSA) (2002) ethical guidelines. Participants were assured that confidentiality would be safeguarded, with anonymity maintained throughout. Identifying features would be removed from the data, with pseudonyms used for all names, including the university itself. We protected the rights of participants by advising that they could withdraw at any time and/or not answer questions. Staff and students were not approached directly but through appropriate office holders. Participants were invited to review the transcripts and data extracts selected for inclusion in the published study. The decision to offer this opportunity recognised and aimed to mitigate the differential power relations between researchers and participants, particularly in the case of undergraduate students.

We reflected throughout on our role as researchers and the associated position of power we occupied. Skeggs (2002: 369) points out that 'telling has always been moral' and reflexivity is needed throughout the process of undertaking research. The data presented here was thematically analysed and reanalysed by both authors. Interview accounts were reflexively constructed and reconstructed through conference papers and research seminars to refine our interpretations and so provide an accurate presentation of the experiences shared with us.

Conceptual Framework

Habitus and Dispositions

This study follows Archer et al. (2012) in drawing on Bourdieu's (1986) concepts, as well as his later insight that family habitus facilitates an individual's 'practical evaluation of the likelihood of the success of a given action in a given situation which brings into play a whole body of wisdom, sayings, commonplaces, (and) ethical precepts (e.g. "that's not for the likes of us")' (1990: 77). Interview transcripts were combed for evidence of dispositions constitutive of habitus, and for examples of

> Unintentional learning made possible by a disposition acquired through domestic or scholastic inculcation of legitimate culture. This transposable disposition, armed with a set of perceptual and evaluative schemes that are available for general application, inclines its owner towards other cultural experiences and enables him to perceive, classify and memorize them differently. (Bourdieu 1984: 20)

In drawing on the theoretical resources of habitus and dispositions, however, the process of acquiring dispositions is viewed as 'durable but not eternal' (Ball 2003: 16). Like Archer et al. (2012), we used these ideas to make sense of the family context and influences operating on individual members as well as to understand their effects.

To theorise the influence of social class background on our participants' choices, we draw on Bourdieu's (1977) concept of habitus and field. We use habitus and field to understand any tensions in the participants' stories between individual agency, shaped by cultural and familial factors, and the wider impact of economic, social and structural factors. We also use the concepts to deepen understanding of the reproduction of class inequalities and of the impact of social class on prospects for intergenerational social mobility (Bourdieu 1986). Habitus is also used to explore the influence of familial dispositions (Crozier et al. 2008).

We analyse social class and its intersections with gender to yield a deeper, more rounded understanding than might arise from a concern with class alone (Brah and Phoenix 2004). Our perspective on social reproduction and inter-generational mobility is informed by Bourdieu's (1986: 46) claim that the social world cannot be understood without recognizing 'capital in all its forms,' including the transmission of economic, social and cultural resources through the family. The term habitus refers to the recurring, durable patterns of social class dispositions and outlook—the beliefs, values, conduct, speech, dress and manners—that are ingrained by everyday experiences within the family, peer group and school (Bourdieu 1979). Habitus conditions and guides individuals as they respond to everyday life without consciously obeying explicit rules. The dispositions to behave in certain ways that constitute habitus are acquired over time and reflect the social conditions within which they were acquired (Mills 2008). The 'social axes of "race"/ethnicity, social class, and gender all contribute to shaping what an individual perceives to be possible and desirable' (Archer et al. 2012: 885).

Science Capital

We draw on the concept of science capital to understand the science-related knowledge, attitudes, experiences and resources that our participants have acquired and how this capital has translated into our participants' engagement with post-18 science (Archer et al. 2015). The concept of science capital is defined by Archer and her colleagues (2015: 928) as 'a conceptual tool for understanding the production of classed patterns in the formation and production of children's science aspirations.' They provide a detailed, if still evolving, conceptual and methodological quantitative tool that can be applied to the investigation of the acquisition and deployment of science capital (Archer et al. 2015). Whilst we do not draw on their work to 'score' science capital we do use the qualitative elements of their conceptualisation of the symbolic value of science to identify and understand science-related resources as

> important contemporary forms of capital, which play a role in the production of social relations of advantage/disadvantage. This is not least because they command a high symbolic and exchange value within contemporary society. (Archer et al. 2015: 926)

Throughout this book we argue that gender and social class remain powerful forces that work subversively to shape and influence the routes and pathways taken by young people. In terms of family background, our data collection approach confirms that intra and intergenerational changes in occupation, income and personal dispositions can be understood only by examining the complex interactions at an individual level that enable some members of each generation to take advantage of available opportunities (Bourdieu 1977).

Social Mobility

Our approach recognises the methodological, conceptual and policy issues involved in understanding the difficulties that hinder social mobility policy, especially since the Coalition Government's explicit drive to increase mobility rates (Wells and Topping 2010: unpaged). We discuss the problematic nature of governmental conceptions of social mobility in Chap. 2 and acknowledge controversies about the extent to which large-scale social mobility is achievable or even desirable.

Our knowledge of social mobility, and of the effectiveness of policies to promote upward movement, is seriously compromised by the multiplicity of variables that contribute to individual and group trajectories and by methodological difficulties in measuring change over time. Social mobility research, for example, has tended to emphasize individual male incomes as an appropriate index of mobility, with father and son earnings compared at selected census points (Lambert et al. 2007). This narrow definition of mobility and the simplistic criteria applied to measure change weaken the value of much research into inter-generational change (Goldthorpe and Jackson 2007). There are also anxieties about the methods used to compare cohorts across the four major studies[1] that have provided so much of the data for social mobility research. We cannot be sure, for example, that social mobility is 'low and falling,' as Blanden et al. (2005b) claim, because only 13 per cent of the National Child Development Study (NCDS) cohort and 12 per cent of the British Cohort Study (BCS) were included in their analysis. This high attrition rate means the difference between the remaining sample members from the 1958 and 1970 cohorts is too small to justify the dramatic claim that social mobility is 'low and falling.' The supposed fall in mobility seems not to have continued for those born from the mid-1980s (Gorard 2008).

Other key issues that bear on our case study investigation of intra- and inter-generational mobility trajectories are discussed in Chap. 2.

Gender

Gender may have consequences for the subject choices some young people make, the forms of assessment they prefer and the sort of career pathway

[1] The four studies are: National Child Development Study (NCDS), from 1958; British Cohort Study (BCS), from 1970; British Household Panel Survey (BHPS), from 1991; Millennium Cohort Study (MCS), from 2000.

they decide to follow. Despite numerous reforms and policy initiatives in the United Kingdom (UK), as elsewhere, for example, Women into Science and Engineering (WISE) (Weiner 1985), gender tends to inform subject choices (Francis et al. 2019). In our work we draw on Berger and Luckmann's (1966) theory of social constructionism and view gender as a concept socially constructed through interaction and discourse.

Burr (2003) argues that studying the social construction of gender requires a critical approach to common sense knowledge and an examination of the historical and cultural specificity of this knowledge. Chilisa (2006: 249) argues that 'gender is a social construction that ascribes roles, behaviours and attitudes on the basis of sex.' By viewing gender as a social construction, we could explore the 'active part that [...] people themselves played in the construction of their gender identities' (Skelton and Francis 2009: 19). This approach to viewing gender has allowed us to challenge the taken for granted norms associated with the gender bias found in some STEM subjects and show the areas where equality policy is working and areas where further improvements are required. It has also enabled us to consider how to challenge the dominant social construction of a scientist.

Chapter Overview

This introduction outlines the study, its design and the key concepts informing our understanding of the ways in which individual agents respond to educational opportunities. The remaining chapters are organised as follows:

In Chap. 2, we review the literature on social mobility and provide a definition of how we understand the concept. We highlight some limitations in existing research on social mobility and examine successive governments' policy interventions aimed at enhancing equal opportunities in terms of gender and social class, and we set out the key reasons why social mobility has acquired such salience in recent decades. In England, Coalition government policies (2010–2015) and subsequent Conservative government policies (2015 to date) to improve opportunity and intragenerational mobility have emphasized the importance of enabling students to access good universities and jobs regardless of social background. Despite the increased funding available for economically disadvantaged pupils through, for example, the pupil premium, the widespread introduction of academies and the adoption of a rigorous curriculum and challenging targets for student outcomes, educational success remains closely associated with relative wealth and social class background. We also draw on literature that builds an international picture in addition to the UK case study to show the similarities and points of difference in approaches to improving participation in STEM subject areas.

Chapter 3 reports on the genealogical case study of 11 families through 3 generations to examine these influences and the operation of underlying social processes. Family histories covering three generations, together with the personal narratives and interpretations of the participants themselves, supply the evidence necessary to assess the extent to which individuals expect to improve their status and potential income relative to other family members. The data also yields a portrait of how participants' aspirations are formed and applied, and shows how these aspirations, as well as their identities and achievements, are related to family and science capital, social class background and education.

In Chap. 4 we draw on staff interview data to examine the chemistry department's equality policies for staff and students. First, there is acknowledgement of the equality progress made in the case study department prior to gaining the ASC. Second, there is discussion of the institutional policies implemented as a consequence of the ASC, including shared parental maternity/paternity leave and unconscious bias interview training. Third, we examine the department's widening participation policy, which aims to improve non-traditional student representation in the department's undergraduate programme. Fourth, there is an examination of the equality initiatives aimed at supporting staff and students, for example, female representation across all areas of work including external speakers, external examiners and student role models.

In Chap. 5 we draw on the interviews with staff to examine areas where the ASC still needs to lead to changes. To what extent and in what ways has the ASC changed the ethos and culture of the department? Are there areas that still require further attention, and if so, how can the ASC fill these gaps? We have identified four key areas where additional forms of support could further enhance equal opportunities. First, improved support for part-time staff applying for promotion is needed. Second, departments should challenge the culture of overworking as the dominant model for achieving career progression. Third, greater representation of Lesbian, Gay, Bisexual, Transgender, Questioning (or queer), and Intersex (LGBTQI+) people, women, ethnic minorities and less advantaged social groups within the staff and student bodies. Fourth, is the need to widen participation to include non-traditional, working class and disadvantaged students.

In Chap. 6 we conclude by reviewing the book's two main purposes. The first is to use genealogical information about family circumstances, education and careers to better understand the formation of young people's identities and aspirations. The second is to review the implementation of chemistry department initiatives at Marsden University that have flowed from work with the Athena Swan Charter, and to consider their effectiveness in the wider context provided by the Equality Act (2010). In reviewing these purposes, we consider our case study data and published research to examine if these policies are convincing or effective. Where they are failing, we consider what can be done to improve the chances of success.

References

AdvanceHE. (2019). AdvanceHE's Athena SWAN Charter. Retrieved February 1, 2020, from https://www.ecu.ac.uk/equality-charters/athena-swan/.

Andrews, J., Robinson, D., & Hutchinson, J. (2017). *Closing the Gap? Trends in Educational Attainment and Disadvantage*. Education Policy Institute. Retrieved June 3, 2019, from https://epi.org.uk/wp-content/uploads/2017/08/Closing-the-Gap_EPI-.pdf.

Archer, L., DeWitt, J., Osborne, J., Dillon, J., Willis, B., & Wong, B. (2012). Science Aspirations, Capital, and Family Habitus: How Families Shape Children's Engagement and Identification with Science. *American Educational Research Journal, 49*(5), 881–908.

Archer, L., Dawson, E., DeWitt, J., Seakins, A., & Wong, B. (2015). "Science Capital": A Conceptual, Methodological, and Empirical Argument for Extending Bourdieusian Notions of Capital Beyond the Arts. *Journal of Research in Science Teaching, 52*(7), 922–948.

Ball, S. (2003). *Class Strategies and the Education Market: The Middle Class and Social Advantage*. London: Routledge Falmer.

Berger, P., & Luckmann, T. (1966). *The Social Construction of Reality: A Treatise in the Sociology of Knowledge*. New York: Random House.

Bertaux, D. (1995). Social Genealogies Commented on and Compared: An Instrument for Observing Social Mobility Processes in the 'Longue Durée'. *Current Sociology, 43*(2), 69–88.

Bertaux, D., & Thompson, P. (Eds.). (1997). *Pathways to Social Class: A Qualitative Approach to Social Mobility*. Oxford: Clarendon Press.

Bhopal, K. (2018). *White Privilege: The Myth of a Post-racial Society*. Bristol: Policy Press.

Blanden, J., Gregg, P., & Machin, S. (2005b). Social Mobility in Britain: Low and Falling. *CentrePiece, 10*, 18–20.
Bourdieu, P. (1977). *Outline of a Theory of Practice*. Cambridge: Cambridge University Press.
Bourdieu, P. (1979). *Algeria 1960: The Disenchantment of the World, the Sense of Honour, the Kabyle House Or the World Reversed*. Cambridge: Cambridge University Press.
Bourdieu, P. (1984). *Distinction*. London: Routledge and Kegan Paul.
Bourdieu, P. (1986). The Forms of Capital. In J. Richardson (Ed.), *Handbook of Theory of Research for the Sociology of Education* (pp. 241–258). New York: Greenwood Press. Translated by R. Nice.
Bourdieu, P. (1990). *Sociology in Question*. Cambridge: Polity Press.
Brah, A., & Phoenix, A. (2004). Ain't I a Woman? Revisiting Intersectionality. *Journal of International Women's Studies, 5*(3), 75–86.
British Educational Research Association (BERA). (2018). Ethical Guidelines for Educational Research. Retrieved December 31, 2018, from http://www.bera.ac.uk/content/ethical-guidelines.
British Sociological Association (BSA). (2002). Statement of Ethical Practice for the British Sociological Association. Retrieved December 31, 2019, from www.britsoc.ac.uk.
Burr, V. (2003). *Social Constructionism* (2nd ed.). East Sussex: Routledge.
Chan, T. W., & Boliver, V. (2013). The Grandparents Effect in Social Mobility: Evidence from British Birth Cohort Studies. *American Sociological Review, 78*(4), 662–678.
Chilisa, B. (2006). Sex Education: Subjugated Discourses and Adolescents' Voice. In C. Skelton, B. Francis, & L. Smulyan (Eds.), *The Sage Handbook of Gender and Education* (pp. 249–261). London: Sage Publications.
Cook, C. (2012, February 22). The Social Mobility Challenge for School Reformers. *FT Data*. Retrieved March 11, 2012, from http://blogs.ft.com/ftdata/2012/02/22/social-mobility-and-schools/#axzz1o06SJkBJ (Registration Required).
Crozier, G., Reay, D., James, D., Jamieson, F., Beedell, P., Hollingworth, S., & Williams, K. (2008). White Middle-Class Parents, Identities, Educational Choice and the Urban Comprehensive School: Dilemmas, Ambivalence and Moral Ambiguity. *British Journal of Sociology of Education, 29*(2), 61–272.
Department for Education (DfE). (2010). *The Importance of Teaching White Paper Equalities Impact Assessment (CM-7980)*. London: The Stationery Office.
Department for Education (DfE). (2017). *Unlocking Talent, Fulfilling Potential: A Plan for Improving Social Mobility Through Education*. Command 9541.
Erola, J., & Moisio, P. (2007). Social Mobility Over Three Generations in Finland 1950–2000. *European Sociological Review, 23*, 169–183.

Fortin, N., Bell, B., & Bohm, M. (2017). Top Earnings Inequality and the Gender Pay Gap: Canada, Sweden, and the United Kingdom. *Labour Economics, 47*, 107–123.

Francis, R., Taylor, R., & Tereshchenko, A. (2019). *Reassessing 'Ability' Grouping: Improving Practice for Equity and Attainment.* London: Routledge.

Friedman, S., & Laurison, D. (2020). *The Class Ceiling: Why It Pays to Be Privileged.* Bristol: Policy Press.

Geertz, C. (1973). *The Interpretation of Cultures: Selected Essays.* New York: Basic Books.

Goldin, C. (2014). Human Capital. In C. Diebolt & M. Haupert (Eds.), *Handbook of Cliometrics* (pp. 55–86). Heidelberg, Germany: Spring-Verlag.

Goldthorpe, J., & Jackson, M. (2007). Intergenerational Class Mobility in Contemporary Britain: Political Concerns and Empirical Findings. *The British Journal of Sociology, 58*(4), 525–546.

Gorard, S. (2008). A Re-consideration of Rates of 'Social Mobility' in Britain: Or Why Research Impact Is Not Always a Good Thing. *The British Journal of Sociology of Education, 29*(3), 317–324.

Hoskins, K., & Barker, B. (2014). *Education and Social Mobility: Dreams of Success.* London: Institute of Education Press.

Hoskins, K., & Barker, B. (2019). Social Mobility: The Potential of a Genealogical Approach. *British Educational Research Journal, 45*(2), 238–253.

Lambert, P., Prandy, K., & Bottero, W. (2007). By Slow Degrees: Two Centuries of Social Reproduction and Mobility in Britain. *Sociological Research Online, 12*(1), 37–62.

Mandler, P. (2017). Educating the Nation: IV. Subject Choice. *Transactions of the Royal Historical Society, 27*, 1–27.

Mare, R. D. (2011). A Multigenerational View of Inequality. *Demography, 48*(1), 1–23.

Mills, C. (2008). Reproduction and Transformation of Inequalities in Schooling: The Transformative Potential of the Theoretical Constructs of Bourdieu. *British Journal of Sociology of Education, 29*(1), 79–89.

Olsen, W., Gash, V., Heuvelman, H., & Walthery, P. (2014). The Gender Pay Gap in the UK Labour Market. In G. Razzu (Ed.), *Gender Inequality in the Labour Market in the UK.* Oxford: Oxford University Press.

Pfeffer, F. (2014). Multigenerational Approaches to Social Mobility. A Multifaceted Research Agenda. *Research in Social Stratification and Mobility, 35*, 1–12.

Powell, A. (2019). Unemployment by Ethnic Background. House of Commons Library. Briefing Paper No. 6385, 22nd May. Retrieved February 9, 2020, from https://researchbriefings.parliament.uk/ResearchBriefing/Summary/SN06385.

Skeggs, B. (2002). Techniques for Telling the Reflexive Self. In T. May (Ed.), *Qualitative Research in Action* (pp. 349–375). London: Sage.

Skelton, C., & Francis, B. (2009). *Feminism and the Schooling Scandal*. London: Routledge.
Smith, E. (2011). Women into Science and Engineering? Gendered Participation in Higher Education STEM Subjects. *British Educational Research Journal, 37*(6), 993–1014.
Social Mobility Commission (SMC). (2016). State of the Nation 2016: Social Mobility in Great Britain. Retrieved April 11, 2018, from www.assets.publishing.service.gov.uk/government/uploads/system/uploads/attachment_data/file/569412/Social_Mobility_2016_Summary_final.pdf.
Thompson, P. (1997). Women, Men, and Transgenerational Family Influences in Social Mobility. In D. Bertaux & P. Thompson (Eds.), *Pathways to Social Class: A Qualitative Approach to Social Mobility* (pp. 32–61). Oxford: Clarendon Press.
Weiner, G. (1985). *Just a Bunch of Girls: Feminist Approaches to Schooling*. Buckingham: Open University Press.
Wells, N., & Topping, A. (2010, August 18). Nick Clegg Attempts to Move Agenda to Social Mobility. *The Guardian*. Retrieved December 3, 2020, from https://www.theguardian.com/politics/2010/aug/18/nick-clegg-social-mobility.
Wilson, H. (1963). *Labour's Plan for Science*. Reprint of Speech by Rt. Hon. Harold Wilson, M.P., Leader of the Labour Party, at the Annual Conference, Scarborough, Tuesday October 1. Retrieved June 9, 2019, from http://nottspolitics.org/wp-content/uploads/2013/06/Labours-Plan-for-science.pdf.

CHAPTER 2

The Contemporary Social Mobility and Equality Policy Context: Framing the Problem

Abstract In Chapter 2, we review the literature on social mobility and provide a definition of how we understand the concept. We highlight some limitations in existing research on social mobility and examine successive governments' policy interventions aimed at enhancing equal opportunities in terms of gender and social class, and we set out the key reasons why social mobility has acquired such salience in recent decades. We also draw on the literature that builds an international picture in addition to the United Kingdom (UK) case study to show the similarities and points of difference in approaches to improving participation in STEM subject areas.

Keywords Social mobility • Policy • Athena Swan Charter • STEM

Introduction

Numerous developments since the 1960s have inspired and conditioned strategies to improve equal opportunities and increase social mobility. This chapter reviews the relevant literature in four key areas to chart and evaluate progress with the meritocratic and mobility policy goals pursued by successive governments and educational organisations. We also review a variety of background circumstances that researchers have identified as

K. Hoskins, B. Barker, *STEM, Social Mobility and Equality*,
https://doi.org/10.1007/978-3-030-49216-8_2

obstacles to progress and present a consequent set of propositions that help us make sense of our case study data and findings. The key areas are

- Science and Education
- Social Mobility
- STEM Initiatives
- Equal Opportunities

Science & Education

Science has been of strategic importance since the late nineteenth century, when the Cavendish laboratories, founded in 1874, played a critical role in the development of modern physics. Although British science could not compete with the breadth, depth and economic relevance of the Kaiser Wilhelm Institutes in Germany (Cornwell 2004), university level research in London, Cambridge and elsewhere was of high quality and contributed to the country's technological strength during the two world wars (Barnett 1986). Scientific research, technology and industry were able to draw on a limited supply of non-elite recruits for a brief period when cities like London, Birmingham and Bristol provided higher grade schools for promising students. These prepared girls as well as boys for the Science and Art Department examinations at age 18. This was an important contribution to scientific and technical education but ceased when the 1902 Act ended their funding (Vlaeminke 1999; Watts 2007). But despite this disappointing decision the potential of science was increasingly recognised. J.B.S. Haldane (1938) was amongst many who popularised scientific discoveries and formed the image of scientific research as a vital source of material progress and optimism about the future. Even so, until the late 1960s there were limited opportunities for young people to study science at school and beyond. Around 80 per cent of children failed the eleven-plus and were excluded from grammar schools and the qualifications necessary to pursue scientific careers (Simon 1991).

A Scientific Revolution

Attitudes and expectations began to change when Harold Wilson, as Leader of the Opposition, chose science as an important campaign theme for the 1964 General Election. *Labour's Plan for Science* (Wilson 1963: 7), with its demand that a new, modern Britain should be 'forged in the white heat' of a scientific revolution, reflected growing concerns prompted by the launch of the Russian Sputnik satellite in 1957. The country was not investing enough in education to meet the demands of parents, never mind those of business and industry; and was failing to equip itself for a modern age when science and technology were vital for economic success and prosperity (Wilson 1963). Wilson's Scarborough speech insisted that more scientists should be produced and that conditions should be created to keep them in the country rather than enlarge the 'brain drain.' Russia, he said, was training 10 to 11 times as many scientists and technologists as Britain. He adopted the arguments of campaigners for comprehensive education to assert that Britain could not afford to waste talent through segregation at age 11 (Wilson 1963; Pedley 1956).

Wilson promised a 'tremendous' building programme for the new universities recommended by the Robbins Report (Martin 2016) and proposed a 'university of the air,' later established as the Open University, to provide opportunities for those who, for one reason or another, had not been able to take advantage of HE (Wilson 1963). The divide in secondary education between grammar and secondary modern schools was increasingly seen as limiting access for working class children and favouring those from middle income backgrounds. The comprehensive school seemed the way to equalise opportunities and so contribute to a fairer and more socially mobile society (Martin 2016). Circular 10/65 requested local education authorities to prepare and submit plans for reorganising on comprehensive lines and Anthony Crosland, Secretary of State for Education (1965–1967), declared the government's intention of ending selection at 11 and eliminating separatism in secondary education (Simon 1991).

Widening Participation

The little acknowledged, long-term outcome of comprehensive reorganisation, later consolidated by the raising of the school leaving age (ROSLA) in 1972 and by the introduction of a unified General Certificate of Secondary Education (GCSE) examination, open to all secondary students in 1986, was an enormous increase in the numbers able to access qualifications, sixth forms and universities (Woodin et al. 2013; Benn and Chitty 1996). These reforms helped create the demand as well as suitable conditions for the expansion of HE. The number of universities almost doubled in the aftermath of John Major's 1992 Further and Higher Education Act; and the following New Labour government announced that 50 per cent of young people should progress to university by 2010

(Blair 2001). Overall participation in HE increased from 3.4 per cent in 1950 to 8.4 per cent in 1970, and from 19.3 per cent in 1990 to 33 per cent by 2000 (Bolton 2012). Total enrolments have increased by 21.4 per cent since 2007–2008 (Universities UK 2018) while overall participation in England reached 49 per cent in 2017 (Adams 2017), with 57.5 per cent of UK university students female and 42 per cent male (Universities UK 2018).

These figures reflect important, sustained changes in education policy that have widened participation and raised the numbers studying in sixth forms and universities. Even so, the underlying policy agenda to improve the UK's economic performance has continued, with a particular emphasis on increasing the flow of skilled, well-qualified recruits into the core science and technology disciplines. This emphasis, potent since *Labour's Plan for Science* in 1963, has intensified with the growth of international competition and the escalating importance of science and technology for business and industry. As Prime Minister, Tony Blair returned to the scientific and educational themes that were so strong in Labour's policymaking in the 1960s:

> The second element is the revolution in technology—and especially information technology. Britain has long been a pioneer in advanced forefront (sic) of technology and scientific development. But science and technology is advancing faster than ever before. For business the key to success is how fast they can adopt and adapt these new technologies. (Blair 2000: unpaged)

Blair, like Harold Wilson before him, saw fairness and equal opportunities as policy priorities that would increase efficiency and reduce the waste of talent associated with disadvantage (McNeil 2012). Between 1997 and 2010, New Labour governments were persuaded that an advanced 'knowledge economy is our best route for success and prosperity' (Blair 2000: unpaged). They responded to the opportunities presented by new technology as well as to perceived shortages in science-related disciplines. Substantial investment and multiple initiatives aimed to improve national competitiveness by increasing the numbers studying STEM subjects and eventually seeking careers in areas of strategic relevance for the economy (Department for Education and Skills [DFES] 2004; Her Majesty's Treasury [HMT] 2006).

Social Mobility

This determined drive to reform education and steer growing numbers of young people towards scientific and technological careers should be understood in the context of New Labour's wider social goals. Policymakers were strongly influenced by human capital theory, and prioritised tackling shortages of suitably trained personnel. But they also wished to create more equal opportunities so young people could overcome disadvantage, achieve vocational success and lead better lives (Gewirtz 2001; McNeil 2012). Gordon Brown, as Chancellor of the Exchequer, was deeply concerned about the long term, damaging consequences of relative poverty and complained about a system that seemed to discourage and even reject students from non-traditional and disadvantaged backgrounds. He insisted it was 'time to end the old Britain where what mattered was the privilege you were born to, not the potential you were born with' (British Broadcasting Corporation [BBC] News 2000: unpaged). Brown's protest seemed all the more justified when a research report suggested that social mobility had fallen for the 1970 birth cohort, compared with the 1958 cohort. Inter-generational income mobility was said to have declined 'over the last few generations of school leavers' (Blanden et al. 2005a: 13).

The report earned great publicity with the statement that social mobility in Britain was 'low and falling' (Blanden et al. 2005b: 18) while the expansion of HE was said to be working more favourably for better off students than for those from poorer backgrounds (Blanden and Machin 2013). Tony Blair interpreted 'low and falling' mobility rates as evidence that institutional bias and poor schools were failing people from non-privileged backgrounds. He said mobility was 'the great force for social equality in dynamic market economies' and his governments promoted social inclusion and education reforms designed to equip young people to compete in the new global economy and achieve upward mobility (Blair 2002: unpaged; McNeil 2012).

Since then, improving access to better jobs has become a pressing international policy concern (Almeida et al. 2012). Initiatives to redress different sources of potential disadvantage, including relative poverty, gender and ethnicity, are now viewed in terms of their possible contribution to increased social mobility (Organisation for Economic Cooperation and Development [OECD] 2019). Social mobility has become the dominant policy theme, with a strong influence on government approaches to a wide range of equality issues. As Secretary of State for Education, for example, Michael Gove stressed the importance of equality (DfE 2010b) but also emphasised the close connection between the moral purpose of education (to enable young people to thrive despite disadvantaged circumstances) and his desired outcome (upward mobility available to all) (Gove 2011).

Growing public debate about the impact of poverty (Jones 2011), increased inequality across the global north and south (Wilkinson and Pickett 2009) and persistent differences in school achievement (Cook 2012; SMC 2016) has strengthened the desire to remove barriers to upward mobility. In the UK, Labour (2003–2010), Coalition (2010–2015) and Conservative (2015 to date) governments have implemented key policies to improve inter- and intra-generational mobility, apparently stuck after a leap forward in employment opportunity in the post-war period (Panel on Fair Access to the Professions [PFAP] 2009; HMG 2009, 2010, 2011; Perkin 1989).

Government interventions have concentrated on barriers that reduce access and opportunity for less advantaged individuals, for those from some ethnic backgrounds and for women and girls constrained by gendered attitudes and practices (Department for Education 2010a, b; HM Government 2011). New Labour insisted on targets and delivery to transform aspirations and results (Barber 2007), while Michael Gove (DfE 2010b) introduced radical reforms to improve school outcomes and increase social mobility. These included raising the quality of teaching and leadership in schools, dramatically extending the Academies programme, reviewing the curriculum to ensure rigour and high standards, and strengthening accountability through a new assessment regime. More recently, the government has published a 'plan for improving social mobility through education' (DfE 2017), with an emphasis on closing the attainment gap between disadvantaged students and the rest. As Secretary of State for Education, Justine Greening said: 'We can break down the barriers that hold people back at every stage in their life …equality of opportunity starts with education' (DfE 2017: 5).

Flaws in Mobility Research

Given the methodological complexity involved in researching mobility, it is no surprise that social mobility studies have yielded contrasting results. One review found eleven studies that report an upward trend, compared with four that have found decline. Another analysis indicates that 50 per cent of the population has changed class position relative to their fathers (Heath and Payne 1999; Lambert et al. 2007). The result is a mixed, imperfect picture of mobility in Britain that provides an inadequate foundation for vast interventions to change underlying social patterns and trends (Erikson and Goldthorpe 2010).

The treatment of women in much mobility research is of particular concern. The emphasis on quantitative studies of male occupations, or income at fixed points in fathers' and sons' lives, presupposes that individual male earnings, rather than those of women and families (Olsen et al. 2014), are the main agent of upward and downward movement (Dex et al. 2009). Women's status is said to derive from male heads of household, in other words, from their fathers and husbands (Dyhouse 2002). As a result, women are deemed to have no independent position (Payne and Abbott 1990). Welfare dependants and others not currently employed are also excluded from consideration (Lambert et al. 2007). The result is that our understanding of the contribution of women and families to social mobility is limited and poorly informed.

The emphasis on male status and incomes in many studies has obscured the enormous growth of female employment in the service sector and also masks the interaction between male and female employment. Mobility analysis should, therefore, include both men and women, with women considered in terms of their own occupation and social class. When this was done for the period from 1992 to 1997, for example, it was clear that, by comparison with their fathers, women continued to benefit from increased opportunity in the expanding service class (Payne and Roberts 2002). The male-centred approach also excludes the role played by the various resources (social, cultural, economic) held by individual men and women but embedded in family structures and transmitted through successive generations (Bourdieu 1986). Property, unearned income and inheritance are now more important for successful households than ever before (e.g. parental support for house purchase, children inheriting houses and capital), and have considerable significance for family members, their social status and their possible mobility (Dorling 2014a, b; Byrne et al. 2017).

Family is an inescapable and important influence on individual career and educational trajectories and helps explain the stability of people's choices and the relative absence of social mobility from lower positions in our previous studies (Sutton Trust 2009; Hoskins and Barker 2014, 2019). Individual agency is to a great extent constrained or enabled by multiple social and economic contexts, with family background, education and labour market fluctuations contributing powerfully to life chances and outcomes (Hoskins and Barker 2017). Current mobility research informs policy unhelpfully because too little attention is paid to families, households, inheritance and capital accumulation (Hoskins and Barker 2019). The BBC Class Survey (BBC Science 2013) is unusual amongst large-scale studies in applying Bourdieu's conception of family resources to the understanding of social differentiation and change, and in recognising the impact of these variables.

Persisting Disadvantage

There is strong evidence that despite efforts to enable young people to overcome their background circumstances, large-scale inequality and disadvantage persist, with profound implications for equal opportunities and social justice (Atkinson 2015). Well-funded schemes to support social inclusion (McNeil 2012), sustained education reforms intended to close the gap between disadvantaged students and their peers (Andrews et al. 2017), and a substantial expansion of HE, have made little impression on young people's life chances (Smith and White 2011). Significant numbers of talented and fortunate individuals may have achieved academic and vocational success, but overall outcomes for disadvantaged groups remain dismal. Background circumstances, including class structures and sizes, continue to exclude many young people from opportunities that appear to be open to all (Payne and Roberts 2002; SMC 2019).

Successive reports show that disadvantage and poor educational outcomes are strongly linked; and that there has been little change over the last twenty years (Social Mobility & Child Poverty Commission 2014; SMC 2016; Andrews et al. 2017). The SMC (2019) reports that by age seven, pupils entitled to free school meals lag behind in reading (18 percentage points), writing (20 percentage points) and mathematics (18 percentage points). Poverty and special educational needs/disabilities are associated, with over 25 per cent of the students eligible for free meals also identified with SEND[1] (Crenna-Jennings 2018). Examination results continue to reflect the distribution of wealth through the social spectrum while those from economically disadvantaged backgrounds are much less likely than their peers to progress to high-status universities, professional

[1] Special Educational Needs and Disability.

careers or well-rewarded jobs (HM Government 2009, 2011; Cook 2012). Those from better off backgrounds are almost 80 per cent more likely to be in a professional job than their working class peers (SMC 2019).

The gap between disadvantaged sixteen-year olds and their peers has been reduced a little since 2007 but these students continue to fall behind better off classmates by approximately 2 months each year over the course of secondary school (Andrews et al. 2017). Comparative analysis of the 1946, 1958 and 1970 birth cohorts shows that parental class, status and education continue to exert a strong influence on academic attainment, even when allowance is made for cognitive ability (Bukodi et al. 2014). The expansion of HE in the UK has disproportionately benefited children from better off backgrounds (Blanden and Machin 2013). Entry to HE is stratified by social characteristics such as occupational class background and economic activity (Smith and White 2011). 22 per cent of the most deprived state school students drop out of university within two years, compared with 7 per cent of the least deprived (Crenna-Jennings 2018).

There is a rhetoric of classlessness but working class and middle-class life patterns remain sharply different, with social class 'everywhere and nowhere, denied yet continually enacted' (Bunting 2007; Reay 2006: 290; Pakulski and Waters 1996). Reay (2018) believes working class children are 'miseducated' and says that 'if you're a working class child, you're starting the race halfway round the track behind the middle class child. Middle class parents do a lot via extra resources and activities' (quoted in Ferguson 2017: unpaged). Families in one study told of extremes of class polarisation around a number of life experiences, confirming the authors' view that social class, gender and race remain powerful in shaping life and life chances (Walkerdine et al. 2001). The class pay gap remains a strong feature of the labour market, and has not changed significantly in recent years. UK professionals from working class backgrounds are paid £6800 less on average each year than those from more affluent families (Sellgren 2017).

There is little evidence that relative rates of class mobility have changed since the 1940s, and it seems that education's main impact has been on the incidence of mobility rather than its rate (Goldthorpe and Mills 2008). Increased numbers participating in HE seem to have consolidated the advantages of some individuals or groups, but the increase in qualifications has had little impact on productivity or skills (Brown 2001). Collins (1981) believes the inflation of credentials has contributed more to competition between status groups (not individuals) than to efficiency and growth.

Flawed Policies

This evidence illustrates the intractable, persistent nature of disadvantage and confirms the pervasive consequences of relative poverty and inequality. But unfounded assumptions and flawed policies have also contributed to these disappointing outcomes. New Labour policy-makers tackled inequalities in housing, child-rearing, education and occupational recruitment but were mistaken in expecting a technological revolution that would increase the demand for graduates and provide great opportunities for social mobility. After 2010, the Conservatives changed the direction of education reform but under-estimated the impact of reductions in financial support for lower income groups, increased university tuition fees and a growing gap between rich and poor (Brown 2013). Austerity created a harsh social climate that exacerbated the plight of the disadvantaged and restricted the scope for upward mobility (O'Hara 2014; Smith 2014).

The SMC (2019) documents the ways in which government decisions since 2010 have reduced rather than increased opportunity for disadvantaged people, with predictable implications for their life chances. Hundreds of children's centres have closed; school funding has been cut by 8 per cent; at age 16 disadvantaged students are more likely to enter further education than school sixth forms but the FE sector is overlooked; in the workplace those with the least skills are the least likely to get training; those from working class backgrounds are more likely to be paid less than the voluntary living wage (SMC 2019). All this is taking place in the context of a 'hollowed out' labour market where the growth in the number of low-level service jobs has been larger than in many other European countries (McIntosh 2013).

There is abundant evidence, therefore, that the policies adopted to improve upward mobility, especially those that emphasise large-scale education reform, are failing to overcome deeply entrenched social and economic barriers to fair access and genuinely equal opportunities (Hoskins and Barker 2019). The supply of talented students from disadvantaged backgrounds seems unlikely to increase without recognition that inequalities in class, status and power load the dice in the struggle for qualifications and high-status jobs; and without appropriate action to narrow social divides (Brown 2013). Education reform alone cannot compensate for fluctuations in the job market, the shape and size of the class structure or the underlying circumstances of people's lives (Barker and Hoskins 2015;

Hoskins and Barker 2019). Governments have simply paid too little attention to the difficulties involved in using education to reduce wider social inequalities and equalise relative life chances (Brown 2013).

The most recent SMC report (2019) is almost despairing in its assessment of progress towards fairness and genuinely equal opportunities for disadvantaged young people. The report is unequivocal in its conclusion that social mobility has 'stagnated over the last four years at virtually all stages from birth to work. Being born privileged in Britain means that you are likely to remain privileged. Being born disadvantaged, however, means that you may have to overcome a series of barriers to ensure that your children are not stuck in the same trap' (SMC 2019: 2). The Commission implores school and university chiefs, politicians and employers to act to ensure that everyone has access to equal opportunities (SMC 2019). But there is less recognition of the need to support disadvantaged young people in ways that enable them to benefit from such opportunities.

STEM Initiatives

This recent evidence suggests that New Labour's vigorous work to promote equal opportunities and increase the flow of socially mobile, highly skilled workers was in many respects flawed and unsuccessful (Gewirtz 2001; Brown 2013) while Coalition and Conservative governments since 2010 have found that austerity and social mobility 'don't mix' (Pickett and Wilkinson 2012: unpaged). Although participation in sixth forms and universities has increased enormously, educational and vocational outcomes continue to follow long established patterns shaped by hierarchy, wealth and family advantage. The introduction of the apprenticeship levy, designed to increase the number of quality and higher-level placements, has been disappointing, with a 24 per cent fall in the number of people starting in-work training. Critics claim low-skill jobs have been rebadged as apprenticeships (BBC 2018). This landscape raises doubts about the likely effectiveness of STEM initiatives designed to influence young people's underlying attitudes and behaviour; and inevitably influences the assessment of policies intended to forge an advanced, scientific and meritocratic society. The desire to create a social order based on scientific progress and merit-based mobility continues to inspire public debate but the meritocratic dream seems remote from current trends.

International Policy Convergence

New Labour's determination to recreate the workforce was shared by many governments around the world. There has been continuous, sustained investment in plans and initiatives to widen participation in science and technology, and to increase national competitiveness in the global economy. These schemes have similar aims, with improvements in science education and other interventions expected to free individuals from the effects of inequality and to encourage them to pursue STEM careers, with significant benefits for themselves and for their employers.

Scientific administrators at the United States (US) National Science Foundation (NSF) introduced the acronym STEM in 2001. Since then, the term has been adopted globally to frame policies designed to increase economic competitiveness through investment in STEM education, training and infrastructure. STEM has become a serious preoccupation for policy-makers, with discussion often driven by perceived shortages of highly skilled labour. Western Europe, Germany, France, Ireland, the Netherlands, Norway, Spain and the UK all have national STEM policies or strategies that provide a coherent framework for greater public knowledge of science, for improving teaching and learning in mathematics and science, and for increasing participation at all levels to ensure an adequate supply of skilled workers for STEM-related business and industry (Marginson et al. 2013).

In the US, the Federal Government has made STEM a top funding priority, with multiple agencies engaged in reforming science and mathematics education (Breiner et al. 2012). By 2005 there were 207 federal education programmes aiming to increase the number of students studying in STEM fields, and also to improve the quality of STEM education (Kuenzi 2008). The working assumption of such policies is that a strong science and innovation system can enable advanced nations to move into high value goods, services and industries, and so compete effectively against low-wage, emerging economies such as China and India (Sainsbury 2007).

In the UK, New Labour assumed that 'the workforce of the future will increasingly require higher-level skills as structural adjustments in the economy force businesses to move up the value chain. These jobs of the future will increasingly require people with the capabilities that a STEM qualification provides' (Targeted Initiative on Science and Maths Education [TISME] 2013: 2). The inevitable conclusion was that more people should study and work in STEM disciplines. But there were mounting concerns that fewer students were choosing STEM subjects. Sir Gareth Roberts review, *Set for Success* (2002), reported significant falls in the number of students taking physics, mathematics, chemistry and engineering qualifications. These downward trends, combined with a shortfall in transferable skills among graduates, seemed to threaten the government's strategy to improve UK productivity and innovation performance.

A 16 per cent drop in entrants to chemistry degrees between 1995 and 2000 was of particular concern (Roberts 2002). The government was alarmed by a marked decline in the numbers choosing to study mathematics, physics and chemistry at A (Advanced) level; and by the falling demand for university places in STEM subjects. There were further worries that women and ethnic minorities were especially under-represented in science and engineering (DFES 2004). This confirmed a widely held conviction that one of the UK's key economic problems was a broad shortage of STEM skills in the workforce (Morse 2018).

Similar concerns and initiatives were reported around the world. After their Lisbon summit in 2000, European Union (EU) leaders pushed the idea of Europe as a competitive and dynamic knowledge-based economy very hard, only to find in 2005 that there were significant shortfalls in the supply of STEM professionals. In this same period there was dismay in the US that the country was producing insufficient numbers of students, teachers and practitioners in the STEM disciplines. Critics believed that a 'large majority' of high school students were failing to achieve proficiency in mathematics and science. Many teachers were said to lack adequate subject knowledge (Kuenzi 2008).

Primary and secondary schools serving disadvantaged communities in the US were said to give inadequate access and exposure to STEM content. As a result, fewer minority students, and female students, were showing interest in STEM-related activity. Lack of diversity in STEM college majors was believed to limit diversity in STEM occupations. Only 9 per cent of STEM freshmen were African-Americans, only 7 per cent were Hispanics and only 1 per cent were Native Americans, compared with 83 per cent who were Caucasian or Asian-American (Avendano et al. 2019).

In Australia it seems that too few young people are coming forward for STEM based education, and that there are worrying rates of attrition for those who embark on STEM courses. This could lead to a dearth of well-qualified graduates equipped for STEM-related employment and contribute to a potential crisis in capacity, of concern to policy-makers across the globe (Bell 2015).

Reported shortfalls prompted UK policy-makers to order a comprehensive mapping of the many initiatives intended to improve the take-up of STEM courses and progression into STEM careers (DFES 2004). The Mapping Review revealed there were over 470 STEM initiatives run by government departments and external agencies, all aiming to encourage students, especially from under-represented groups, to get involved with STEM subjects at school (Smith 2010). The review noted that while there was a major effort in primary and secondary schools there was little activity in further and higher education.

Among the successes was the Wider Horizons initiative that won an Institute of Careers Guidance Award in 2005. This showed there was considerable potential to expand work experience by including stimulating placements to raise aspirations and widen access to STEM careers. Sheffield Hallam University initiated an enhanced work experience scheme for girls, working with the local Education Business Partnership, local schools and employers. Preparation and support were provided to reduce stereotypical placement choices (Morton and Collins 2010).

The Mapping Review was followed by the publication of the Ten-Year Science and Innovation Investment framework (SIIF), a long-term plan to ensure a strong and successful UK research base, designed to span the years 2004–2014 (HMT 2006). The Office of Science and Innovation (OSI) was established within the Department of Trade & Industry (DTI) to bring responsibility for science and innovation in central government together for the first time. Its role was to manage UK science policy and ensure funding for research, to be allocated via the Research Councils. The SIIF included proposals to increase the supply of STEM skills (HMT 2006).

The continued emphasis on science and innovation has given STEM subjects a privileged position in the UK government's HE policy. They are seen as strategically important and have been protected in a climate of general austerity and budget reductions (Smith and White 2017). The 2010–2015 Parliament provided £185 million to underpin the teaching of high-cost STEM subjects in HE, and contributed £7.2 million to support science teachers through the National Science Learning Network between 2014 and 2016 (Smith and White 2017). The Department for Children, Schools and Families (DCSF) played an important role in developing a better delivery system for STEM initiatives and activities (Tripney et al. 2010).

These measures in the UK have coincided with a major expansion of HE, with the number of candidates accepted for first degrees quadrupling in twenty years (Smith 2011). This overall growth has been accompanied by much higher female rates of participation, with women 7.2 per cent ahead of men by 2006. Similar increases in female participation have been observed in most OECD countries, with the exception of Japan, Korea, Turkey, Germany and Mexico (Broecke and Hamed 2008). In the UK, overall growth is fully reflected in science and science-related subjects where the numbers studying for full-time undergraduate degrees have trebled since 1986, with 200,000 new entrants in 2009 (Smith 2011).

In the US, large sums have been invested in pursuit of similar goals. The National Institutes of Health (NIH) received $998 million in 2004 to fund 51 programmes to increase participation and quality in STEM education; and the National Science Foundation (NSF) received $997 million to pay for 48 programmes (Kuenzi 2008). Policy-makers have been determined to maintain the country's scientific and technological lead by ensuring a continuous supply of skilled STEM students and workers.

STEM strategies in the UK have been reinforced by school reforms to raise standards (DfE 2010a), and multiple initiatives to increase successful participation by under-represented groups, including disadvantaged students, women and girls, and ethnic minorities (Her Majesty's Government [HMG] 2011). The public benefits of education and training to improve the nation's human capital appear to match the private benefits for individuals who gain access to an expanding supply of high-quality jobs and enjoy more equal opportunities (Smith 2011). Science, technology, economic growth, equal opportunities and social mobility seemed to belong to a virtuous and progressive circle.

Shortages

There are reasons, nevertheless, to question some of the assumptions that have driven policy. Perceived and reported shortages of suitably qualified personnel have prompted many countries to emphasise STEM education but doubts have been expressed about the nature and extent of the supposed skills crisis. Resource consultants suggest that 'The skill shortage might just be a convenient myth for employers, when in fact they don't understand what job seekers really want' (ARM 2016: 2). Others see a link between economic hard times and calls for improved STEM education. In Australia, for example, the economic depressions of the 1890s, 1930s and 1980s have a clear correlation with significant developments in technology education. The financial crisis of 2007–2009 may be part of a pattern, with economic difficulties prompting an emphasis on scientific and technical education (Williams 2011).

Recent analysis of data from the Higher Education Statistical Agency (HESA), covering more than three million UK graduates from 1994 to 2010, confirms these doubts. STEM graduates fare little better than non-STEM graduates in the job market. Graduates in some STEM subjects do better than average, but those qualified in other STEM disciplines do less well than those with non-STEM degrees. The findings of the analysis seem incompatible with a true shortage of potential STEM workers (Smith and White 2017). Smith and White (2018b) are confident that in the UK there is no overall shortage of STEM graduates but concede there may be a shortage of STEM graduates willing to work in highly skilled STEM roles in some occupational areas, and that employers may experience a shortage of suitable candidates.

Smith and White (2018b) conclude nevertheless that the usual indicators associated with labour shortages, for example, very low levels of unemployment, are not present in their data. They also point to the remarkable fact that most STEM graduates never work in highly skilled STEM jobs. STEM graduates are no more likely to enter graduate positions than those with degrees in other subjects and are just as likely to be unemployed. There is no evidence that STEM workers have an advantage in the labour market of the kind that might be expected if there were a genuine problem with supply. The STEM subjects have become the holy grail of twenty-first century education in England, with female students outnumbering males in A-level science entries for the first time, but there is a risk that the government's drive to steer young people away from the arts and humanities towards science and technology may be misconceived (Weale and Larsson 2019).

Widening Participation and Unequal Opportunities

There are also strong indications that STEM initiatives have failed so far to broaden the recruitment base for science subjects. The main patterns of participation in STEM and STEM-related subjects amongst UK-based undergraduates have been relatively stable in terms of students' backgrounds and ethnic group. Policies have widened access but there has been minimal impact on unequal academic and vocational outcomes for young people from less advantaged backgrounds. There is also little sign of movement towards greater equality in participation rates amongst advantaged and disadvantaged undergraduates in general. The occupational class distribution of science students resembles that of their peers and their pattern of participation in HE is similar (Smith 2011).

Government statistics on widening participation confirm the extent to which inequality persists and social background continues to exert a substantial influence on destinations and outcomes. An estimated 25.6 per cent of pupils in receipt of Free School Meals (FSM) at age 15 entered HE by age 19 in 2016/17. This compares with 43.3 per cent of non-FSM pupils achieving the same goal. The gap in entry rates between FSM and non-FSM pupils has remained at 17.7 per cent over the last three years; and has varied between 16.8 per cent and 19.2 per cent since 2005/06. There is an even larger gap (26.7 per cent) between the rates achieved by students from the most disadvantaged geographical areas (20.4 per cent) and those from the most advantaged areas (47.1 per cent) (DfE 2018). Another indicator of persisting inequality is provided by the student progression rate, from A levels and equivalent qualifications to the most selective HE providers. The rate for independent school and college students is 61.4 per cent; the rate for state-funded schools and colleges is 22.4 per cent. The gap in progression rates has fluctuated between 37.4 per cent in 2008/09 and 39 per cent in 2016/17 (DfE 2018). The expansion of HE seems to have made the socio-economic gap in university participation worse rather than better (Fumagalli 2019).

School level STEM interventions to increase participation in 'pure' science subjects like chemistry and physics also cause concern because their impact on long-term recruitment at Advanced (A) level has been limited. Many initiatives have had little or no effect. Fewer male students are following physics courses than fifty years ago and the numbers have been in decline since the 1990s. There has been no significant change in the number of females studying the subject (Smith 2011). The number of UK students applying to study chemistry at university, stable for at least a decade, fell by 21 per cent between 2015 and 2018, despite an increase in young people sitting A level in the subject (Royal Society of Chemistry, quoted in McKie 2019: unpaged).

Concerned drug companies have sent an open letter to the government calling for urgent action to encourage school leavers and students to follow careers in chemistry (McKie 2019). Once a male-dominated subject in sixth forms and universities, however, chemistry has become gender-neutral, with the participation gap between men and women progressively narrowed from the early 1990s, mainly due to a decline in male numbers (Smith 2011). Science became compulsory with the introduction of the National Curriculum in 1988 but making students study science subjects to age 16 has not increased candidate numbers at A level and beyond. Female participation in 'pure' science has proved stable, despite strong, well-funded campaigns to encourage positive attitudes and engage girls in STEM activities. There are many more women studying science and science-related subjects but female recruitment to physics and engineering remains relatively small. There has been no corresponding increase in the numbers studying biology (Smith 2011; Fowler 2019).

Women, Science & Work

The overall picture of women's place in the job market, however, is much more positive and challenges the perception that female candidates are less interested or less able in science. Girls do not lack ambition and now comprise 60 per cent of all medical undergraduates and 75 per cent of those taking veterinary science (Macdonald 2014). For at least twenty years, approximately 30 per cent of female entrants to HE have chosen science or science-related disciplines. Women are fully engaged with science and wish to study the subject at the highest levels; about a half of science-related places in HE are taken by women (Smith 2011). After university, female STEM graduates are just as likely to be employed in graduate jobs (86 per cent of females compared with 87 per cent of males) but they are less likely to work in the high skill STEM sector (32 per cent female compared with 55 per cent male) (Smith and White 2018a).

There is a similar picture in the US, with the American Association of University Women reporting that high school girls and boys perform equally well in maths and science. The majority of college graduates in the US are female (57 per cent) and women are well represented in life sciences, chemistry and mathematics, although they account for only 20 per cent of bachelor's degrees in engineering, computer science and physics (Girl Scouts 2012). In Australia, women provide about 80 per cent of commencers in health and slightly more than half in agriculture and the environment, and in the natural and physical sciences. As in the UK and US, women in Australia are much less well represented in engineering and related technologies (20 per cent) and in information technology (14 per cent) (Marginson et al. 2013).

Policy-makers generally perceive a problem with female under-representation in areas they consider strategic and important (i.e. physics, engineering and computer science) but the bias in women's choices (e.g. towards medicine and health-related areas, psychology) is not necessarily problematic and suggests well-grounded concerns and identities not easily aligned with government priorities (Smith 2011). The employment data is less encouraging, however. The large and sustained increase in the number and quality[2] of female graduates (Thompson and Bekhradnia 2009; HESA 2013) has not yet produced genuinely equal opportunities in the work-place for women and men. A recent measure of the median full-time pay gap indicates a differential of 9.4 per cent (BITC 2012), and most female options and careers continue to be limited by gendered constraints (Francis 2006) or are simply gendered. Women have increased their share of managerial and professional jobs but relatively few have made it to the top. Although one in four women are junior managers, only 1 or 2 per cent join senior management. Only 12 per cent of partners in law firms are female; only 15 per cent of medical consultants are female; and only 9 per cent of university professors are women. Women's progress towards achieving senior and professional posts does not reflect their outstanding academic achievement (Walkerdine et al. 2001).

Greenfield et al.'s report on Women in STEM (2002) concluded that women's progress was hindered by traditional family roles, responsibilities and expectations, with unequal pay and intangible cultural factors excluding them from the corridors of power. The report argued that the gender pay gap reinforced problems with work/life balance, attitudes towards family responsibilities and the lack of childcare. In addition, a gendered hierarchy was encoded in organisational culture as images, metaphors, artefacts, beliefs, values, norms, rituals, language, legends and myths in ways that constituted institutionalised sexism (Greenfield et al. 2002).

Support was not readily available for women struggling to balance childcare with research careers, leading to the unequal opportunities that were responsible for the high male/female ratio at senior levels. This bias persisted although men and women were equally represented as junior scientists. Asked whether opportunities had improved ten years later, Greenfield responded that nothing much had changed and would not

[2] As measured by degree class.

change 'until people put much more money into maternity and childcare schemes, until they have crèches in much higher abundance near laboratories' (Attar 2012: unpaged).

There are continued concerns about the under-representation of women in STEM careers and the 'leaky' pipeline that is apparent as women drop out at every stage in the STEM journey. Even when fully engaged with science careers, women may not be encouraged to pursue higher-level jobs by their employers and may struggle with time constraints as they enter the childbearing years (Fowler 2019: unpaged). Decisions about gaining qualifications, setting up an independent home, having children and taking care-breaks all have immense and far-reaching impact on women's lives (Walkerdine et al. 2001).

Various initiatives have tackled the problem of under-representation. In the UK, the Engineering Council collaborated with the Equal Opportunities Commission in 1984 to launch the WISE campaign. Their mission was to get one million more women active in the STEM workforce, to boost the talent pool and to drive performance and growth for employers and the UK economy (WISE 2014). The Girls into Science and Technology (GIST) initiative also aimed to increase female participation in STEM careers (Smail et al. 1982). The Centre for Innovation in STEM education at California State University established the Women in STEM Education (WISE) programme in the US, with a similar mission to increase female participation through mentorships, leadership opportunities, field trips and STEM activities (Avendano et al. 2019).

Despite these efforts, the striking gender gap in the proportion of students taking maths and physics to A level persists. The most recent data (2017) shows that 18 per cent of girls, compared with 33 per cent of boys, took maths A level, and only 4 per cent of girls compared with 17 per cent of boys took physics A level. The gap in chemistry in 2017 was just 2 percentage points. The data cannot be explained by gender differences in prior attainment. Low confidence, especially in physics, appears to contribute to girls' reluctance to take maths and physics A level, and a majority of respondents in a recent study agreed that these subjects are male dominated. The Institute of Fiscal Studies found that women were also discontented with the style of maths and physics teaching at school. Focus groups highlighted girls' concerns about excessive content, a boring and repetitive emphasis on examination preparation and limited opportunities to go into depth on topics they found interesting (Cassidy et al. 2019).

Explaining Aspirations

Government policies and initiatives, intended to improve equal opportunities, to encourage social mobility and to widen participation in STEM-related subjects, have encountered numerous interacting resistances that have impeded progress. Policy-makers have failed to recognise 'sociology's inconvenient truth' (Brown 2013: 678); and have under-estimated or misunderstood the influence of women and families on young people's choices and trajectories (Hoskins and Barker 2014, 2019). They have denied the relevance and impact of social class and inequality (Savage et al. 2013; Wilkinson and Pickett 2009); and have failed to examine how recession, austerity and welfare reform have contributed to mental health problems and widening inequalities (O'Hara 2014; Barr et al. 2015). Policy problems and solutions have been treated as if they were much more straightforward than they are in practice (Gove 2012; DfE 2010a, b, 2017).

TISME, funded by the Economic and Social Research Council (ESRC) in partnership with the Institute of Physics, the Gatsby Foundation and the Association for Science Education (ASE), represents a serious and sustained attempt to promote a more informed approach, through a careful investigation of the complex social and educational territory that has to be understood before STEM initiatives can succeed. Led by Louise Archer at King's College, London, TISME aims to map young people's aspirations and how they engage with science and mathematics education, as well as to encourage greater participation, achievement and understanding (KCL 2019).

Important insights have been derived from TISME's *Aspires* project, a five-year, longitudinal study of how children develop science and career aspirations, using a national survey of children aged 10–14. Although young people like science and express interest in further study, many decide not to because they have narrow ideas about the usefulness of science qualifications. They do not feel 'clever' enough to pursue post-16 science and STEM-related careers. The survey has also shown that families are an important influence on young people's STEM aspirations. This is consistent with Holman and Finegold's (2010) finding that young people regard families and friends as the most trusted source of careers advice.

Children with a close relative working in a science-related career are much more likely to continue with science in the sixth form than their peers, and they are also more likely to pursue STEM-related employment.

Science capital, defined in terms of a family's stock of science-related qualifications, interests, understanding and social contacts, was found to have a significant but not determining role in leading young people to be interested in STEM and to have a positive view of scientific pathways (TISME 2013).

Habitus & Dispositions

The project's analysis and interpretation of survey and interview evidence builds a conceptual framework to make sense of differences and inequalities in students' responses to science in the early years of secondary education (age 10–14). The team uses Bourdieu's (1986) concepts to help us understand the ways in which the interplay of social inequalities in the distribution of capital[3] and differences in family habitus combine to produce uneven class and race-based patterns in children's science aspirations; and also to shape their attitudes toward future participation in science careers (Archer et al. 2012).

The *Aspires* research uses the concept of family habitus to consider how families build a relationship with science and explores how this is influenced by their possession of the various forms of capital. The survey found that students from affluent, professional and middle-class backgrounds, with high levels of cultural capital, described more positive parental attitudes to science. Powerful effects were produced by the inter-action of habitus and capital in families with science specific capital; and by a strongly pro-science habitus that included child-rearing practices known as 'concerted cultivation' (Lareau 2003). The family context made science familiar, accessible and desirable, as well as providing children with additional resources. Interest and even careers in science were spread through the family and could often be traced back over several generations, with parents and other individuals holding science degrees and/or working in science-related occupations (Archer et al. 2012).

Middle class families that lacked science specific capital were nevertheless more likely to possess the resources to enable and support a variety of science activities. Working class families, by contrast, were found to be constrained by the interplay of family habitus and the unequal distribution of both generic and science specific capital. It seems that 'for a sizeable

[3] The term capital is used to denote resources that can be economic, cultural, social and symbolic.

proportion of working class families and children, science is simply an "unthinkable" aspiration' (Archer et al. 2012: 899) and that social class is of great importance 'in facilitating or constraining children's potential science aspirations and identifications' (Archer et al. 2012: 903). This was found to be true although the overwhelming majority of children in the sample said they liked science.

The *Aspires* data, from over 18,000 students across three surveys, provides a convincing picture (from year 6 to year 8) of students' responses to science-related subjects and learning:

> A student is most likely to express science aspirations if he is male, Asian, has high/very high levels of cultural capital, is in the top set for science and has a family member who works in science or a STEM-related job. A student is least likely to see science as 'for me' if she is female, White, has low/very low levels of cultural capital, is in the bottom set and does not have any family members who use science in their jobs. (Archer et al. 2013: 3)

This survey data seems to show that class and inequality are woven into young people's dispositions and point towards their social reproduction (Bourdieu and Passeron 1977). Some young people are confirmed in their scientific aspirations, while others believe scientific studies and careers are 'unthinkable' for 'people like us.' There is some evidence from the survey that students can also exercise personal agency in choosing science despite lack of knowledge and resources at home. Bourdieu's ideas need not predetermine children's choices (Archer et al. 2013: 3).

Gendered Aspirations

This interpretation is consistent with discouraging evidence, cited earlier, that inequality and disadvantage work to undermine equal opportunities and produce profoundly different responses to apparently open and accessible lessons and school experiences.

The *Aspires* survey also captures responses that help explain the gender bias and gender gaps detailed above:

> Gender issues are evident from a young age. Girls are less likely than boys to aspire to science careers, even though a higher percentage of girls than boys rate science as their favourite subject. Girls are far more likely to aspire to arts-related and 'caring' careers. Among 12–13-year-old students, 18 per

> cent of boys and 12 per cent of girls aspire to become scientists—in comparison, 64 per cent of girls aspire to careers in the arts. (Archer et al. 2013: 3)

The qualitative survey data shows that girls who are keen about science are likely to come from middle-class backgrounds and have access to abundant science capital (Archer et al. 2013).

Equalities

The *Aspires* research illustrates how the possession of resources shapes family habitus and influences dispositions that relay and reproduce social class and inequality. Once acquired, dispositions are not easily changed and have a strong influence on behaviour. Policies to reconstruct the workforce through a virtuous circle of science, technology and apparently equal opportunities have failed to reduce these class effects. On the contrary, there is strong evidence (SMC 2019) that formally equal opportunities have yielded seriously unequal outcomes. New class relations have emerged over time, not least through struggles against the inequalities implicit in neoliberal workplace reforms, but these are marked by increased rather than reduced inequality (Wilkinson and Pickett 2009; Tyler 2015). Hard-won mobility for a talented or fortunate few cannot compensate for the persisting 'legacies of disadvantage' (Bloome 2014: 1197) that stem from inequality and resist individualist solutions. Strongly unequal social structures seem to constrain students' perceptions of what is possible for them and to inhibit their thoughts about STEM subjects and careers (Wilkinson and Pickett 2009; Archer et al. 2012).

The effects of inequality and social exclusion, especially from decent housing (Goulden 2016), have been under-estimated and also compounded by what Owen Jones (2011: 269) describes as 'the demonization of the working class.' A 'discourse of derision' (Wallace 1993) that began with the popular press and an attack on progressive education has been extended to include class, gender, race, ethnic minorities as well as the Lesbian, Gay, Bi-Sexual and Transgender (LGBT) community. An example of the racist references and stereotypes that have seeped into public writing is provided by the current Prime Minister's novel *Seventy Two Virgins*, said to refer to Arabs as having 'hook noses' and 'slanty eyes'; to a mixed-race person as 'coffee-coloured' or 'half-caste'; and to a woman as 'a mega-titted six-footer' (Sandhu 2019). These attempts to ridicule and marginalise increase the sense of injustice felt by minority groups, and prompt activists to challenge negative stereotypes and demand recognition, equal rights and equal treatment. Long struggles against discrimination and injustice are driven by the need to assert people's identities, affirm their worth and secure fairness (Kahn and Sveinsson 2015). They also apply pressure on policy-makers to give greater attention to equality issues with the potential to unlock the 'wasted talent' that has troubled us since the 1960s (Jones 2011; Tyler 2015).

Equal Rights

Campaigners have long recognised that equal rights and equal treatment are not easily secured, even when evidence of discrimination and injustice is overwhelming; and that apparent success, in the form of legislative change, major shifts in public opinion and affirmative action to redress unfair practice, may fail to provide lasting relief (Keck 2009; Bloome 2014). Progress is uneven and can be reversed. The suffragette movement is an early, modern example of the power of well-organised campaigning, but illustrates the sheer difficulty of moving beyond recognition and formal legal rights towards full and lasting equality in ordinary life (Joannou and Purvis 1998). Successive feminist waves have transformed the rights, opportunities and intellectual resources available to women nevertheless, not least by arguing that characteristics such as race, class, gender, sexuality, disability and age do not operate as discrete entities but intersect and combine to position individuals differentially (Strelitz 2004; Collins and Bilge 2016).

In certain optimistic times and places (e.g. 1960s in Britain and the US), groups have been successful in lobbying governments and legislators for political support and changes in the law to secure recognition and greater fairness. In less favourable circumstances (e.g. after the 2008 crash), activists have adopted the techniques of the perpetual campaign to draw attention to their cause and promote their aims (Adonis 2013; Dryden 2018). Disability Rights UK (2012), for example, provides a handbook for campaigners and aims to lower the gap in employment between disabled people and non-disabled people, and to 'truly tackle hate crime' (2012: unpaged) by encouraging an increase in the number of reported crimes and incidents. Disability is a protected characteristic under the 2010 Equality Act and disabled people have received significant financial support over time. But effective campaigning and legal protection have proved of limited use as austerity has deepened, with disability cuts now said to be starving the sick (O'Hara 2014; Ryan 2018).

LGBT campaigns have won remarkable changes in public attitudes and the law (Dryden 2018). There is a government LGBT action plan to 'strip away the barriers that hold people back' with promised action on hate crime and bullying (Government Equalities Office 2018: 1). But gay conversion therapy is still legal, there have been extended school gate protests about children being taught that LGBT people exist, and 24 per cent of homeless young people are from an LGBT background (Ashenden and Parsons 2019). Higher Education Institutions (HEIs) aim to be inclusive but the Equality Challenge Unit (ECU) reports a high incidence of homophobic behaviour, including derisory comments or homophobic banter (ECU 2010). Advance can provoke backlash, with governments themselves prone to prejudice and irrational action (Keck 2009). Section 28 of the 1988 Local Government Act, for example, prevented local authorities from promoting homosexuality or 'pretended family relationships' and prohibited them from funding relevant educational materials and projects (Dryden 2018: unpaged).

Race relations legislation has also failed to guarantee equal or fair treatment for oppressed ethnic minorities. Twice as many people from Black, Asian and Minority Ethnic (BAME) backgrounds, for example, are unemployed compared with those from a White background, although there is substantial variation between different ethnic minority groups. Among people aged 16–24, unemployment rates are highest for those from a Black background (26 per cent) compared with a rate of 11 per cent for young people from a White background (Powell 2019). Black workers with degrees earn over 23 per cent less on average than White workers with similar qualifications (Equality and Human Rights Commission [EHRC] 2016). Disproportionate numbers of working age adults from minority groups have incomes below 60 per cent of the median and experience a higher prevalence of child deprivation (Darton and Strelitz 2004). Black people in England are three times more likely to be murdered and are four times more likely to be stopped and searched by the police (EHRC 2016).

Equality Act, 2010

Entrenched and persistent injustice in various areas of social life has prompted equality specialists, human rights organisations and other campaigners to lobby for changes in the law and for the strengthening of bodies like the Equal Opportunities Commission and the Commission for Racial Equality, set up to undertake strategic enforcement and to assist individuals. The resulting Equality Act (2010) replaced nine previous major pieces of legislation; established the EHRC; seeks to implement fully four main EU directives; and outlaws numerous forms of discrimination. The Act adopts a unitary or integrated perspective of equality law enforced by a single commission; clarifies the definitions of discrimination, harassment and victimisation and applies them across the protected characteristics;[4] expands the positive duties on public authorities to advance equality in respect of all the protected characteristics; widens the circumstances in which positive action is allowed; and creates a new duty on public authorities to have due regard to socio-economic disadvantage when taking strategic decisions (Hepple 2010). Class background is not a protected characteristic, however, and governments since 2010 have declined to make the 'socio-economic duty' defined in the Act (section "Introduction") effective (The Equality Trust 2018).

After a long campaign and much pressure, the principal aim of the 2010 Act is to harmonise, simplify and modernise equality law. The legislation is based on the principle that equality is an indivisible fundamental human right and that there can be no hierarchy of equality.

There are weaknesses in this approach. Although a remedy is provided for intersectional discrimination that may arise from combinations of the protected characteristics, the essential logic of the Equality Act (2010) is for redress to be sought for grievances in relation to particular protected characteristics, on an additive basis if necessary. The Equality Act brings together all the relevant statutes but retains the single dimension logic of separate discrimination claims. This leads to a loss of synergy between the characteristics (Solanke 2011). The concept of stigma should be introduced instead, it is argued, because the phrase 'stigmatized characteristics' would remind us that 'equality law is designed to address oppression rather than just promote diversity' (Solanke 2011: 352).

Athena Swan

These strenuous efforts to assert equal rights in ways that ensure equal access and equal opportunity for everyone, regardless of their background and personal characteristics, provide the context for this case study in

[4] See section "Introduction".

HE. How far can HEIs overcome known resistances to equality? To what extent can HEIs develop their organisation and internal processes so that all their students can succeed and feel safe and comfortable in so doing? How effective are the ASC and the Race Equality Charter (REC) in facilitating the work of HEIs to achieve their objectives?

The ASC developed from work between the Athena Project and the Scientific Women's Academic Network (SWAN) that aimed to increase the representation of women in science, technology, engineering, medicine and mathematics (STEMM) (AdvanceHE 2019). The ECU, now known as AdvanceHE and owned by Universities UK, has been the leading partner in advancing equality in further and higher education and assists institutions build a culture that values diversity and equality (Bhopal 2018). Its charters (ASC and REC) aim to encourage and drive forward cultural and systemic changes that make a real impact on the lives of staff and students (AdvanceHE 2019).

First introduced in 2005, the ASC's agreed principles were:

- To address gender inequalities requires commitment and action from everyone at all levels of the organisation
- To tackle the unequal representation of women in science requires changing culture and attitudes
- The absence of diversity at management and policy-making levels has broad implications which the organisation will examine
- The high loss rate of women in science is an urgent concern which the organisation will address
- The system of short-term contracts has particularly negative consequences for the retention and progression of women in science, which the organisation recognises
- There are both personal and structural obstacles to women making the transition from Doctor of Philosophy (PhD) into a sustainable academic career in science, which require the active consideration of the organisation.

Institutions have to confirm their commitment to these principles to participate. Eligible HEIs or departments within them may apply for Bronze, Silver and Gold Charter Awards.

In 2010 the ECU identified a pressing need to prioritise race equality and began to develop a specific charter on the model of the ASC. After consultation and a trial with volunteer institutions, the REC was fully launched in 2016. The REC aims to improve the representation, progression and success of minority ethnic staff and students within HE. There are five guiding principles:

- Racial inequalities are a significant issue within HE; racism is an everyday facet of UK society
- UK HE cannot reach its full potential until individuals from all ethnic backgrounds can benefit equally from the opportunities available
- Solutions should aim for long-term institutional culture change and avoid solutions aimed at individuals
- The complexity of different backgrounds, experiences and outcomes should be considered in analysing data and shaping solutions
- Individuals have multiple identities; the intersection of different identities should be considered whenever possible

The ASC itself was extended in May 2015 to recognise work in arts, humanities, social sciences, business and law, and in professional and support roles, and for trans staff and students. The ASC now covers women (and men where appropriate) in academic roles, professional and support staff, and trans staff and students. It is concerned with people's representation, the progression of students into academia, their journey through career milestones and the working environment for all staff. There are 164 ASC members, holding 815 awards between them (AdvanceHE 2019).

In 2013 the ECU commissioned a research team from Loughborough University to examine the impact of the Charter in HEIs. This found that 90 per cent of the institutional champions and the vast majority of departmental champions (81 per cent) agreed that ASC had impacted positively on gender issues. Most champions said there had been a positive impact on women's career progression. Survey data and qualitative findings confirm that the ASC process and award is considered to be of great value. One institutional champion commented: 'It's the most effective standard/process/lever for change I've come across in 12 years of equality work, including impact assessment' (Munir et al. 2014: 9).

Critics see the ASC as a positive move to tackle gender inequalities but argue that by its nature and application to STEM, it has become 'an example of the privileging of whiteness and white identity' (Bhopal 2018: 52) because the main beneficiaries have been mainly middle class and white. Black and Minority Ethnic (BME) academic staff are less likely to be in senior positions, while BME women are said to be the most disadvantaged in HE. They are less likely to be professors and face greater barriers to promotion than their white colleagues (Bhopal 2018).

Another review of ASC was launched in May 2018 to ensure the Charter's rigour and credibility. The Steering Group's update has noted ASC's strengths. The Charter is said to have had a positive influence and to be valued by the sector. The process has facilitated conversations about gender equality at all levels across institutions; has identified areas for action and has led to changes in institutional policy; has driven changes in culture and ethos; and has provided additional support and developmental opportunities for staff (Buckingham 2019).

Our investigation of a highly successful science department (Chap. 4) aims to evaluate the progress that has been made towards Charter criteria; and to consider how far the Charter process offers a convincing solution to the resistances of class, gender and ethnicity that have undermined so many previous initiatives to create equal rights and opportunities. We conclude this chapter by identifying five propositions implicit in the literature that seem to explain why that resistance is so strong. Does the case study confirm the propositions or are there signs of a breakthrough that can be generalised to support equalities work more widely?

Propositions

1. Inequality is a stronger influence than formal government policies;
2. Family background, capital and inequality are powerful influences on expectations, choices and pathways;
3. Social class, gender and ethnicity have a major influence on self-identity and aspirations;
4. These identity influences are stronger than equal opportunities policy and its effects, and explain why the pattern documented by Smith and White (2011) persists despite much apparently worthy policy activity;
5. The weakness of the Equality Act 2010 stems from its protected nine characteristics because they do not include social class and also fail to recognise adequately the intersectionality of multiple identities and the synergy between the protected characteristics.

References

Adams, R. (2017, September 28). Almost Half of All Young People in England Go on to Higher Education. *The Guardian*. Retrieved May 1, 2019, from https://www.theguardian.com/education/2017/sep/28/almost-half-of-all-young-people-in-england-go-on-to-higher-education.

Adonis, A. (2013, January 4). How Labour Can Learn from Roy Jenkins. *The Guardian*. Retrieved May 1, 2019, from https://www.theguardian.com/commentisfree/2013/jan/04/how-labour-can-learn-roy-jenkins.

Advanced Resource Managers (ARM). (2016). *The Truth Behind The STEM Skill Shortage: A Serious Mismatch Between Hiring Company Perceptions and Job Seeking Reality*. ARM White Paper, June. Retrieved June 3, 2019, from https://www.arm.co.uk/media/1645/arm-the-truth-behind-the-stem-skill-shortage.pdf.

AdvanceHE. (2019). AdvanceHE's Athena SWAN Charter. Retrieved February 1, 2020, from https://www.ecu.ac.uk/equality-charters/athena-swan/.

Almeida, R., Arbelaez, J., Honorati, M., Kuddo, A., Lohmann, T., Ovadiya, M., Pop, L., Puerta, M., & Weber, M. (2012). *Improving Access to Jobs and Earnings Opportunities: The Role of Activation and Graduation Policies in Developing Countries*. Social Protection and Labor Discussion Paper No. 1204. Retrieved June 3, 2019, from http://siteresources.worldbank.org/SOCIALPROTECTION/Resources/SP-Discussion-papers/430578-1331508552354/1204.pdf.

Andrews, J., Robinson, D., & Hutchinson, J. (2017). *Closing the Gap? Trends in Educational Attainment and Disadvantage*. Education Policy Institute. Retrieved June 3, 2019, from https://epi.org.uk/wp-content/uploads/2017/08/Closing-the-Gap_EPI-.pdf.

Archer, L., DeWitt, J., Osborne, J., Dillon, J., Willis, B., & Wong, B. (2012). Science Aspirations, Capital, and Family Habitus: How Families Shape Children's Engagement and Identification with Science. *American Educational Research Journal, 49*(5), 881–908.

Archer, L., Osborne, J., DeWitt, J., Dillon, J., Wong, B., & Willis, B. (2013). Aspires: Young People's Science and Career Aspirations, Age 10–14, Department of Education and Professional Studies, King's College, London. Funded by the ESRC, Research Grant Number RES 179-25-0008, as Part of the Targeted Initiative on Science and Mathematics Education (TISME). Retrieved August 28, 2015, from https://www.kcl.ac.uk/ecs/research/aspires/aspires-final-report-december-2013.pdf.

Ashenden, A., & Parsons, V. (2019, June 13). 15 Rights LGBT People in the UK Still Don't Have. *Pink News*. Retrieved June 3, 2019, from https://www.pinknews.co.uk/2019/06/13/15-rights-lgbt-uk-still-dont-have-pride/.

Atkinson, A. (2015). *Inequality. What Can Be Done?* Cambridge, MA: Harvard University Press.

Attar, N. (2012). *Susan Greenfield on Women in Science: Not Much Has Improved in 10 Years.* BMC/Springer, Blog Network. Retrieved September 1, 2018, from https://blogs.biomedcentral.com/on-biology/2012/03/08/susan-greenfield-on-women-in-science-not-much-has-improved-in-10-years/.

Avendano, L., Renteria, J., Kwon, S., & Hamdan, K. (2019). Bringing Equity to Underserved Communities Through STEM Education: Implications for Leadership Development. *Journal of Educational Administration and History, 51*(1), 66–82.

Barber, M. (2007). *Instruction to Deliver.* London: Politico's Publishing.

Barker, B., & Hoskins, K. (2015). Can High-Performing Academies Overcome Family Background and Improve Social Mobility? *British Journal of Sociology of Education, 38*(2), 221–240.

Barnett, C. (1986). *The Audit of War.* Basingstoke and Oxford: Macmillan.

Barr, B., Kinderman, P., & Whitehead, M. (2015). Trends in Mental Health Inequalities in England During a Period of Recession, Austerity and Welfare Reform 2004 to 2013. *Social Science and Medicine, 147*, 324–331.

Bell, D. (2015). Perceptions of STEM, Within the Context of Teaching Design and Technology in Secondary Schools; A Phenomenographic Study. Edge Hill University. Retrieved May 1, 2019, from https://repository.edgehill.ac.uk/6331/1/.

Benn, C., & Chitty, C. (1996). *Thirty Years On: Is Comprehensive Education Alive and Well Or Struggling to Survive?* London: David Fulton Publishers, Ltd.

Bhopal, K. (2018). *White Privilege: The Myth of a Post-racial Society.* Bristol: Policy Press.

Blair, T. (2000, March 7). Tony Blair's Full Speech, 'The e-generation is with us'. *The Guardian.* Retrieved July 20, 2016, from https://www.theguardian.com/uk/2000/mar/07/tonyblair.

Blair, T. (2001, May 23). Full Text of Tony Blair's Speech on Education. *The Guardian.* Retrieved July 20, 2016, from https://www.theguardian.com/politics/2001/may/23/labour.tonyblair.

Blair, T. (2002, June 10). Full Text of Tony Blair's Speech on Welfare Reform. *The Guardian.* Retrieved July 20, 2016, from http://www.guardian.co.uk/society/2002/jun/10/socialexclusion.politics.

Blanden, J., & Machin, S. (2013). Educational Inequality and the Expansion of United Kingdom Higher Education. *Scottish Journal of Political Economy, 60*, 5.

Blanden, J., Gregg, P., & Machin, S. (2005a). *Intergenerational Mobility in Europe and North America.* London: Centre for Economic Performance, London School of Economics.

Blanden, J., Gregg, P., & Machin, S. (2005b). Social Mobility in Britain: Low and Falling. *CentrePiece, 10*, 18–20.

Bloome, D. (2014). Racial Inequality Trends and the Intergenerational Persistence of Income and Family Structure. *American Sociological Review, 79*(6), 1196–1225.

Bolton, P. (2012). *Education: Historical Statistics*. House of Commons Library. Retrieved July 20, 2016, from https://dera.ioe.ac.uk//22771/.

Bourdieu, P. (1986). The Forms of Capital. In J. Richardson (Ed.), *Handbook of Theory of Research for the Sociology of Education* (pp. 241–258). New York: Greenwood Press. Translated by R. Nice.

Bourdieu, P., & Passeron, J.-C. (1977). *Reproduction in Education, Society and Culture*. London: Sage.

Breiner, J., Harkness, S., Johnson, C., & Koehler, C. (2012). What Is STEM? A Discussion about Conceptions of STEM in Education and Partnerships. *School Science and Mathematics, 112*(1), 3–11.

British Broadcasting Corporation (BBC) News. (2000). Chancellor Attacks Oxford Admissions. Retrieved March 23, 2018, from http://news.bbc.co.uk/1/hi/education/764141.stm.

British Broadcasting Corporation (BBC) News. (2013). The Great British Class Calculator: What Class Are You? Retrieved June 3, 2018, from www.bbc.co.uk/news/magazine-22000973.

British Broadcasting Corporation (BBC) News Business. (2018, April 19). Apprenticeships Down Amid 'Frustration'. Retrieved May 15, 2019, from https://www.bbc.co.uk/news/business-43809679.

Broecke, S., & Hamed, J. (2008). *Gender Gaps in Higher Education Participation: An Analysis of the Relationship Between Prior Attainment and Young Participation by Gender, Socio-Economic Class and Ethnicity*. DIUS Research Report 08 14, Department for Innovation, Universities and Skills. Retrieved July 15, 2015, from http://webarchive.nationalarchives.gov.uk/+/http:/www.dius.gov.uk/research/documents/DIUS-RR-08-14.pdf.

Brown, D. (2001). The Social Sources of Educational Credentialism: Status Cultures, Labor Markets, and Organizations. *Sociology of Education, 74* (Extra Issue: Current of Thought: Sociology of Education at the Dawn of the 21st Century), 19–34.

Brown, P. (2013). Education, Opportunity and The Prospects for Social Mobility. *British Journal of Sociology of Education, 34*(5–6), 678–700.

Buckingham, J. (2019). Athena Swan Review – Update from the Steering Group, AdvanceHE. Retrieved January 1, 2020, from https://www.advance-he.ac.uk/news-and-views/Athena%20SWAN%20Review%20-%20update%20from%20the%20Steering%20Group.

Bukodi, E., Erikson, R., & Goldthorpe, J. (2014). The Effects of Social Origins and Cognitive Ability on Educational Attainment: Evidence from Britain and Sweden. *Acta Sociologica, 57*(4), 293–310.

Bunting, M. (2007, January 24). The Class Debate Demands to Be Heard. *The Guardian*. Retrieved December 31, 2019, from https://www.theguardian.com/commentisfree/2007/jan/24/society.comment.

Business in the Community (BITC). (2012). *Women and Work: The Facts*. Fact Sheet. Retrieved July 20, 2016, from http://opportunitynow.bitc.org.uk/research-insight/WomenWorkFactsheet.

Byrne, A., Shlomo, N., & Chandola, T. (2017). Multilevel Modelling Approach to Analysing Life Course Socioeconomic Status and Compensating for Missingness. Retrieved July 12, 2019, from http://hummedia.manchester.ac.uk/institutes/cmist/archive-publications/working-papers/2017/MLM%20approach%20to%20analysing%20life%20course%20SES%20and%20compensating%20for%20missingness.pdf.

Cassidy, R., Cattan, S., Crawford, C., & Dytham, S. (2019). How Can We Increase Girls' Uptake of Maths and Physics A Level? Institute for Fiscal Studies. Retrieved December 31, 2019, from https://www.ifs.org.uk/uploads/R149%20with%20cover.1.pdf.

Collins, R. (1981). Crises and Declines in Credentialing Systems. In R. Collins (Ed.), *Sociology Since Mid-Century* (pp. 191–215). New York: Academic Press.

Collins, P., & Bilge, S. (2016). *Intersectionality*. Cambridge: Polity Press.

Cook, C. (2012, February 22). The Social Mobility Challenge for School Reformers. *FT Data*. Retrieved March 11, 2012, from http://blogs.ft.com/ftdata/2012/02/22/social-mobility-and-schools/#axzz1o06SJkBJ (Registration Required).

Cornwell, J. (2004). *Hitler's Scientists: Science, War and the Devil's Pact*. London: Penguin Books.

Crenna-Jennings, W. (2018). *Education in England: Annual Report*. Education Policy Institute. Retrieved April 11, 2019, from https://epi.org.uk/wp-content/uploads/2018/07/EPI-Annual-Report-2018-Lit-review.pdf.

Darton, D., & Strelitz, J. (Eds.). (2004). *Tackling UK Poverty and Disadvantage in the Twenty-First Century: An Exploration of the Issues*. York: Joseph Rowntree Foundation.

Department for Education (DfE). (2010a). *The Importance of Teaching: The Schools White Paper*. London: The Stationery Office.

Department for Education (DfE). (2010b). *The Importance of Teaching White Paper Equalities Impact Assessment (CM-7980)*. London: The Stationery Office.

Department for Education (DfE). (2017). *Unlocking Talent, Fulfilling Potential: A Plan for Improving Social Mobility Through Education*. Command 9541.

Department for Education (DfE). (2018). *Widening Participation in Higher Education, England, 2016/17 Age Cohort – Official Statistics*. GOV.UK.

Department for Education and Skills (DFES). (2004). Report on the Science, Technology, Engineering and Maths (STEM) Mapping Review, Volume 1, Main Report, Higher Education Directorate.

Dex, S., Bukodi, E., & Joshi, H. (2009, December 8). *The Ups and Downs of Men's and Women's Careers*. Institute of Education, GeNet Seminar, London. Retrieved July 14, 2015, from http://www.cls.ioe.ac.uk.

Disability Rights UK. (2012). Campaigns. Retrieved July 1, 2015, from https://www.disabilityrightsuk.org/policy-campaigns/campaigns.

Dorling, D. (2014a). *Inequality and the 1%*. London: Verso Books.

Dorling, D. (2014b). Thinking About Class. *Sociology, 48*(3), 452–462.

Dryden, S. (2018). *A Short History of LGBT Rights in the UK*. British Library. Retrieved December 31, 2019, from https://www.bl.uk/lgbtq-histories/articles/a-short-history-of-lgbt-rights-in-the-uk.

Dyhouse, C. (2002). Graduates, Mothers and Graduate Mothers: Family Investment in Higher Education in Twentieth-Century England. *Gender and Education, 14*(4), 325–333.

Equality and Human Rights Commission (EHRC). (2016). Healing a Divided Britain: The Need for a Comprehensive Race Equality Strategy. Retrieved December 31, 2019, from https://www.equalityhumanrights.com/en/publication-download/healing-divided-britain-need-comprehensive-race-equality-strategy.

Equality Challenge Unit (ECU). (2010). Advancing LGB Equality: Improving the Experience of Lesbian, Gay and Bisexual Staff and Students in Higher Education. Retrieved June 4, 2019, from www.ecu.ac.uk/publications/advancing-lgb-equality.

Erikson, R., & Goldthorpe, J. (2010). Has Social Mobility in Britain Decreased? Reconciling Divergent Findings on Income and Class Mobility. *The British Journal of Sociology, 61*(2), 211–230.

Ferguson, D. (2017, November 21). Working-Class Children Get Less of Everything in Education – Including Respect. *The Guardian*. Retrieved November 21, 2017, from https://www.theguardian.com/education/2017/nov/21/english-class-system-shaped-in-schools.

Fowler, S. (2019, January 10). Women in Stem: Critical to Innovation. *Global Policy Journal*: Wiley-Blackwell.

Francis, B. (2006). Troubling Trajectories: Gendered "Choices" and Pathways from School to Work. In C. Leathwood & B. Francis (Eds.), *Gender and Lifelong Learning: Critical Feminist Engagements* (pp. 57–69). London: Routledge.

Fumagalli, L. (2019). Short and Long Term Barriers to University Education in England. What Do Expectation Data Tell Us? ISER: University of Essex. Retrieved February 1, 2020, from https://www.iser.essex.ac.uk/files/projects/information-expectations-transitions-he/barriers-university-education-england.pdf.

Gewirtz, S. (2001). Cloning the Blairs: New Labour's Programme for the Resocialization of Working-Class Parents. *Journal of Education Policy, 16*(4), 365–378.

Girl Scouts of the USA. (2012). *Generation Stem: What Girls Say About Science, Technology, Engineering, and Maths.* Report from the Girl Scout Research Institute. Retrieved October 25, 2018, from https://www.girlscouts.org/join/educators/generation_stem_full_report.pdf.

Goldthorpe, J., & Mills, C. (2008). Trends in Intergenerational Class Mobility in Modern Britain: Evidence from National Surveys, 1972–2005. *National Institute Economic Review, 205*(1), 83–100.

Goulden, C. (2016). *There Can Be No Social Mobility Without More Affordable Housing.* Joseph Rowntree Foundation. Retrieved November 19, 2017, from https://www.jrf.org.uk/blog/there-can-be-no-social-mobility-without-more-affordable-housing.

Gove M. (2011). The Moral Purpose of School Reform, Address at the National College for School Leadership, Birmingham, 16 June, Unpaged. Retrieved March 30, 2018, from http://www.education.gov.uk/inthenews/speeches/a0077859/the-moral-purpose-ofschool-reform.

Gove, M. (2012). A Coalition for Good – How We Can All Work Together to Make Opportunity More Equal, Speech at Brighton College, 10 May, Unpaged. Retrieved March 30, 2018, from http://www.education.gov.uk/inthenews/speeches/a00208822/brightoncollege.

Government Equalities Office. (2018). *LGTB Action Plan: Improving the Lives of Lesbian, Gay, Bi-Sexual and Transgender People.* London: HMSO.

Greenfield, S., Peters, J., Lane, N., Rees, T., & Samuels, G. (2002). *Set Fair, A Report on Women in Science, Engineering, and Technology.* London: Department of Trade and Industry.

Haldane, J. (1938). *Science and Everyday Life.* Allahabad. First Indian Edition 1945: India Publishers.

Heath, A., & Payne, C. (1999). *Twentieth Century Trend in Social Mobility in Britain.* Working Paper No. 70, Centre for Research into Elections and Social Trends. Retrieved July 15, 2015, from http://www.crest.ox.ac.uk/papers/p70.pdf.

Hepple, B. (2010). The New Single Equality Act in Britain. *The Equal Rights Review, 5*, 11–24.

Higher Education Statistics Agency. (2013). Student Introduction 2012/13. Retrieved July 15, 2015, from https://www.hesa.ac.uk/intros/stuintro1213.

HM Government (HMG). (2009). *New Opportunities: Fair Chances for the Future.* Norwich: HMSO.

HM Government (HMG). (2010). *Unleashing Aspiration: The Government Response to the Final Report of the Panel on Fair Access to the Professions.* London: HMSO.

HM Government (HMG). (2011). *Opening Doors, Breaking Barriers: A Strategy for Social Mobility.* London: Cabinet Office.
HM Treasury (HMT). (2006). *Science and Innovation Investment Framework 2004–2014: Annual Report.* London: HM Treasury.
Holman, J., & Finegold, P. (2010). *Report to the Gatsby Charitable Foundation.* Warwick University. Retrieved April 15, 2018, from https://warwick.ac.uk/fac/soc/ier/ngrf/stem/movingon/research/500-stem_careers_review_nov_2010_holman.pdf.
Hoskins, K., & Barker, B. (2014). *Education and Social Mobility: Dreams of Success.* London: Institute of Education Press.
Hoskins, K., & Barker, B. (2017). Aspirations and Young People's Constructions of Their Futures: Investigating Social Mobility and Social Reproduction. *British Journal of Educational Studies, 65*(1), 45–67.
Hoskins, K., & Barker, B. (2019). Social Mobility: The Potential of a Genealogical Approach. *British Educational Research Journal, 45*(2), 238–253.
Joannou, M., & Purvis, J. (Eds.). (1998). *The Women's Suffrage Movement: New Feminist Perspectives.* Manchester: Manchester University Press.
Jones, O. (2011). *Chavs: The Demonization of the Working Class.* London: Verso Books.
Kahn, O., & Sveinsson, K. (Eds.). (2015). *Race and Elections.* Runnymede Perspectives, Runnymede Trust. London: LSE.
Keck, T. (2009). Beyond Backlash: Assessing the Impact of Judicial Decisions on LGBT Rights. *Law & Society Review, 43*(1), 151–186.
Kings College London (KCL). (2019). TISME – Targeted Initiative on Science and Maths Education. Retrieved May 3, 2019, from https://www.kcl.ac.uk/ecs/research/research-centres/cppr/research/pastproj/tisme/index.
Kuenzi, J. (2008). Science, Technology, Engineering, and Mathematics (STEM) Education: Background, Federal Policy, and Legislative Action, Congressional Research Service Reports. 35. Retrieved October 2, 2019, from http://digitalcommons.unl.edu/crsdocs/35.
Lambert, P., Prandy, K., & Bottero, W. (2007). By Slow Degrees: Two Centuries of Social Reproduction and Mobility in Britain. *Sociological Research Online, 12*(1), 37–62.
Lareau, A. (2003). *Unequal Childhoods: Class, Race, and Family Life.* Berkeley: University of California Press.
Macdonald, A. (2014). *Not for People Like Me? Under-Represented Groups in Science, Technology and Engineering.* South East Physics Network: WISE. Retrieved December 31, 2019, from https://www.wisecampaign.org.uk/wp-content/uploads/2018/06/not_for_people_like_me-full-report.pdf.
Marginson, S., Tytler, R., Freeman, B., & Roberts, K. (2013). STEM: Country Comparisons: International Comparisons of Science, Technology, Engineering and Mathematics (STEM) Education. Final Report. Australian Council of

Learned Academies, Melbourne, Victoria. Retrieved June 3, 2018, from http://dro.deakin.edu.au/view/DU:30059041.

Martin, J. (2016). Education Policy. In A. Crines & K. Hickson (Eds.), *Harold Wilson: The Unprincipled Prime Minister – Reappraising Harold Wilson* (pp. 131–148). London: Biteback Publishing.

McIntosh, S. (2013). *Hollowing Out and the Future of the Labour Market.* BIS Research Paper Number 134. Department for Business, Innovation and Skills (DBIS). Retrieved September 18, 2016, from https://assets.publishing.service.gov.uk/government/uploads/system/uploads/attachment_data/file/250206/bis-13-1213-hollowing-out-and-future-of-the-labour-market.pdf.

McKie, R. (2019, August 31). Fears for Britain's Standing in World of Science as Students Shun Chemistry Degrees. *The Guardian.* Retrieved November 25, 2019, from https://www.theguardian.com/education/2019/aug/31/students-shun-chemistry-degrees-university-applications-fall.

McNeil, C. (2012). *The Politics of Disadvantage: New Labour, Social Exclusion and Post-crash Britain.* Institute for Public Policy Research (IPPR): Lankelly Chase Foundation. Retrieved September 19, 2016, from https://lankellychase.org.uk/wp-content/uploads/2016/05/IPPR_Report-Politics-of-Disadvantage-20121213.pdf.

Morse, A. (2018). *Delivering STEM (Science, Technology, Engineering and Mathematics) Skills for the Economy.* Report by the Comptroller and Auditor General National Audit Office, Department for Business, Energy & Industrial Strategy, and Department for Education.

Morton, P., & Collins, J. (2010). Widening Horizons in Science, Technology, Engineering and Maths (STEM) Work Experience. Retrieved January 19, 2019, from https://warwick.ac.uk/fac/soc/ier/ngrf/stem/digging/equality/mortonandcolins_wideninghorizonsinstemworkexperience.pdf.

Munir, F., Mason, C., McDermott, H., Morris, J., Bagilhole, B., & Nevill, M. (2014). Evaluating the Effectiveness and Impact of the Athena SWAN Charter, Executive Summary, Equality Challenge Unit, Athena SWAN, Loughborough University. Retrieved March 3, 2019, from https://www.ecu.ac.uk/wp-content/uploads/external/evaluating-the-effectiveness-and-impact-of-the-athena-swan-charter.pdf.

O'Hara, M. (2014). *Austerity Bites: A Journey to the Sharp End of Cuts in the UK.* Bristol: Polity Press.

Olsen, W., Gash, V., Heuvelman, H., & Walthery, P. (2014). The Gender Pay Gap in the UK Labour Market. In G. Razzu (Ed.), *Gender Inequality in the Labour Market in the UK.* Oxford: Oxford University Press.

Organisation for Economic Co-operation and Development (OECD). (2019). *Action Needed to Tackle Stalled Social Mobility.* OECD Home. Retrieved

January 9, 2020, from https://www.oecd.org/newsroom/action-needed-to-tackle-stalled-social-mobility.htm.

Pakulski, J., & Waters, M. (1996). *The Death of Class*. London: Sage Publications.

Panel on Fair Access to the Professions (PFAP). (2009). *Unleashing Aspiration: The Final Report of the Panel on Fair Access to the Professions*. London: Cabinet Office.

Payne, G., & Abbott, P. (Eds.). (1990). *The Social Mobility of Women: Beyond Male Mobility Models*. Basingstoke: The Falmer Press.

Payne, G., & Roberts, J. (2002). Opening and Closing the Gates: Recent Developments in Male Social Mobility in Britain. *Sociological Research Online, 6*, 4, unpaged.

Pedley, R. (1956). *Comprehensive Education: A New Approach*. London: Victor Gollancz Ltd.

Perkin, H. (1989). *The Rise of Professional Society*. London: Routledge.

Pickett, K., & Wilkinson, R. (2012, May 15). Sorry Nick Clegg – Social Mobility and Austerity Just Don't Mix. *The Guardian*. Retrieved July 18, 2018, from https://www.theguardian.com/commentisfree/2012/may/15/nick-clegg-social-mobility-austerity.

Powell, A. (2019). Unemployment by Ethnic Background. House of Commons Library. Briefing Paper No. 6385, 22nd May. Retrieved February 9, 2020, from https://researchbriefings.parliament.uk/ResearchBriefing/Summary/SN06385.

Reay, D. (2006). The Zombie Stalking English Schools: Social Class and Educational Inequality. *British Journal of Educational Studies, 54*(3), 288–307.

Roberts, G. (2002). *Set for Success: The Supply of People with Science, Technology, Engineering and Mathematics Skills*. Report of Sir Gareth Roberts' Review, HM Treasury. Retrieved August 12, 2016, from https://web.archive.org/web/20070205114712/http://www.hmtreasury.gov.uk/media/643/FB/ACF11FD.pdf.

Ryan, F. (2018, April 5). A Year of Disability Cuts Has Done Nothing But Starve the Sick. *The Guardian*. Retrieved November 19, 2019, from https://www.theguardian.com/commentisfree/2018/apr/05/year-disability-cuts-starving-sick-employment.

Sainsbury, D. (2007). *The Race to the Top: A Review of Government's Science and Innovation Policies*. HM Treasury: HMSO.

Sandhu, S. (2019, November 28). Boris Johnson's Novel Seventy-Two Virgins Criticized for Racist and Sexist Language, Inews. Retrieved December 6, 2019, from https://inews.co.uk/news/politics/boris-johnson-novel-seventy-two-virgins-racist-sexist-language-1327014.

Savage, M., Devine, F., Cunningham, N., Taylor, M., Li, Y., Hjellbrekke, J., et al. (2013). A New Model of Social Class? Findings from the BBC's Great British Class Survey Experiment. *Sociology, 47*, 219–250.

Sellgren, K. (2017, January 26). *Social Mobility: Class Pay Gap Found in UK Professions*. BBC News, Family & Education. Retrieved June 17, 2018, from www.bbc.co.uk/news/education-38744122.

Simon, B. (1991). *Education and the Social Order 1940–1990*. London: Lawrence & Wishart.

Smail, B., Whyte, J., & Kelly, A. (1982). Girls into Science and Technology: The First Two Years. *School Science Review, 63*, 620–629.

Smith, E. (2010). Do We Need More Scientists? A Long-Term View of Patterns of Participation in UK Undergraduate Science Programmes. *Cambridge Journal of Education, 40*(3), 281–298.

Smith, E. (2011). Women into Science and Engineering? Gendered Participation in Higher Education STEM Subjects. *British Educational Research Journal, 37*(6), 993–1014.

Smith, M. (2014). *The Impact of Austerity on Schools and Children's Education and Well-Being*. The Encyclopaedia of Informal Education. Retrieved July 3, 2018, from http://infed.org/mobi/the-impact-of-austerity-on-schools-and-childrens-education-and-well-being/.

Smith, E., & White, P. (2011). Who Is Studying Science? The Impact of Widening Participation Policies on the Social Composition of UK Undergraduate Science Programmes. *Journal of Education Policy, 26*(5), 677–699.

Smith, E., & White, P. (2017). A 'Great Way to Get On'? The Early Career Destinations of Science, Technology, Engineering and Mathematics Graduates. *Research Papers in Education, 32*(2), 231–253.

Smith, E., & White, P. (2018a). *The Employment Trajectories of Science Technology Engineering and Mathematics Graduates, Final Report*. Leicester: University of Leicester/University of Warwick.

Smith, E., & White, P. (2018b). Where Do All the STEM Graduates Go? Higher Education, the Labour Market and Career Trajectories in the UK. *Journal of Science Education and Technology, 28*, 26–40.

Social Mobility and Child Poverty Commission (SMCPC). (2014). State of the Nation 2014: Social Mobility and Child Poverty in Great Britain, Report Summary. Retrieved July 16, 2017, from https://assets.publishing.service.gov.uk/government/uploads/system/uploads/attachment_data/file/367461/State_of_the_Nation_-_summary_document.pdf.

Social Mobility Commission (SMC). (2016). State of the Nation 2016: Social Mobility in Great Britain. Retrieved April 11, 2018, from www.assets.publishing.service.gov.uk/government/uploads/system/uploads/attachment_data/file/569412/Social_Mobility_2016_Summary_final.pdf.

Social Mobility Commission (SMC). (2019). State of the Nation 2018–19: Social Mobility in Great Britain. Retrieved December 12, 2019, from https://assets.publishing.service.gov.uk/government/uploads/system/uploads/attachment_data/file/798687/SMC_State_of_Nation_2018-19_Summary.pdf.

Solanke, I. (2011). Infusing the Silos in the Equality Act 2010 with Synergy. *Industrial Law Journal, 40*(4), 336–358.

Strelitz, J. (2004). Tackling Disadvantage: Education. In D. Darton & J. Strelitz (Eds.), *Tackling UK Poverty and Disadvantage in the Twenty-First Century An Exploration of the Issues* (pp. 57–75). London: Joseph Rowntree Foundation.

Sutton Trust. (2009). *Introduction in Social Mobility and Education: Academic Papers Presented at a High Level Summit Sponsored by the Carnegie Corporation of New York and the Sutton Trust, 2008*. Putney: Sutton Trust. Retrieved June 6, 2018, from https://www.suttontrust.com/research-paper/social-mobility-education/.

Targeted Initiative on Science and Mathematics Education (TISME). (2013). What Influences Participation in Science and Mathematics? Retrieved August 28, 2015, from https://www.kcl.ac.uk/sspp/departments/education/research/aspires/TISME-briefing-paper-March-2013.pdf.

The Equality Trust. (2018). Socio-Economic Duty Unpaged. Retrieved December 6, 2019, from https://www.equalitytrust.org.uk/socio-economic-duty.

Thompson, J., & Bekhradnia, B. (2009). Male and Female Participation and Progression in Higher Education. Full Report, Higher Education Policy Institute (HEPI) No. 41. Retrieved July 14, 2015, from http://www.hepi.ac.uk/2009/06/05/male-and-female-participation-and-progression-in-higher-education/.

Tripney, J., Newman, M., Bangpan, M., Niza, C., MacKintosh, M., & Sinclair, J. (2010). *Factors Influencing Young People (Aged 14–19) in Education About STEM Subject Choices: A Systematic Review of the UK Literature*. Wellcome Trust, EPPI Centre, Social Science Research Unit: Institute of Education, University of London.

Tyler, I. (2015). Classificatory Struggles: Class, Culture and Inequality in Neoliberal Times. *The Sociological Review, 63*, 493–511.

Universities UK (UUK). (2018). *Patterns and Trends in UK Higher Education*. London: Universities UK. Retrieved January 3, 2020 https://www.universitiesuk.ac.uk/facts-and-stats/data-and-analysis/Documents/patterns-and-trends-in-uk-higher-education-2018.pdf.

Vlaeminke, M. (1999). *The English Higher Grade Schools: A Lost Opportunity*. London: Woburn Press.

Walkerdine, V., Lucey, H., & Melody, J. (2001). *Growing Up Girl: Psychosocial Explorations of Gender and Class*. Basingstoke: Palgrave.

Wallace, M. (1993). Discourse of Derision: The Role of the Mass Media Within the Education Policy Process. *Journal of Education Policy, 8*(4), 321–337.

Watts, R. (2007). *Women in Science: A Social and Cultural History*. Abingdon: Routledge.

Weale, S., & Larsson, L. (2019, August 16). 'Students Don't See the Value': Why A-Level English Is in Decline. *The Guardian*. Retrieved December 3, 2020, from https://www.theguardian.com/education/2019/aug/16/students-dont-see-value-a-level-english-decline.

Wilkinson, R., & Pickett, K. (2009). *The Spirit Level: Why Equality Is Better for Everyone*. Harmondsworth: Penguin Books.

Williams, J. (2011). STEM Education: Proceed with Caution. *Design and Technology Education, 16*(1), 26–35.

Wilson, H. (1963). *Labour's Plan for Science*. Reprint of Speech by Rt. Hon. Harold Wilson, M.P., Leader of the Labour Party, at the Annual Conference, Scarborough, Tuesday October 1. Retrieved June 9, 2019, from http://nottspolitics.org/wp-content/uploads/2013/06/Labours-Plan-for-science.pdf.

Women into Science Technology, Engineering and Mathematics (WISE). (2014). *The Talent Pipeline from Classroom to Boardroom*. UK Statistics 2014. WISE: Campaign for Gender Balance in Science, Technology & Engineering. Retrieved December 3, 2019, from https://wisecampaign.scdn3.secure.raxcdn.com/wp-content/uploads/2018/04/WISE_UK_Statistics_2014.pdf.

Woodin, T., McCulloch, G., & Cowan, S. (2013). *Secondary Education and the Raising of the School-Leaving Age: Coming of Age?* London: Palgrave Macmillan.

CHAPTER 3

Making Chemists

Abstract Chapter 3 reports on the genealogical case study of eleven families through three generations to examine these influences and the operation of underlying social processes. Family histories covering three generations, together with the personal narratives and interpretations of the participants themselves, supply the evidence necessary to assess the extent to which individuals expect to improve their status and potential income relative to other family members. The data also yields a portrait of how participants' aspirations are formed and applied, and shows how these aspirations, as well as their identities and achievements, are related to family and science capital, social class background and education.

Keywords Family background • Science capital • Genealogical • Aspirations

Introduction

A genealogical approach, as explained in Chap. 1, provides a relevant, valid method for investigating how far family networks, habitus and dispositions influence participants in building science capital and deciding to study chemistry at a Russell Group university. To what extent is family a vital element in their success? Are the propositions (1), (2) and (3) identified earlier in Chap. 2: 48 valid and justified?

K. Hoskins, B. Barker, *STEM, Social Mobility and Equality*,
https://doi.org/10.1007/978-3-030-49216-8_3

This chapter adopts, therefore, a genealogical case study of eleven families through three generations to examine these influences and the operation of underlying social processes. These family histories, together with the personal narratives and interpretations of the participants themselves, supply the evidence necessary to assess the extent to which individuals expect to improve their status and potential income relative to other family members. The data also yields a portrait of how participants' aspirations are formed and applied, and shows how these aspirations, as well as their identities and achievements, are related to family and science capital, social class background and education (Table 3.1).

Seven key themes emerged from analysis of the 11 interview transcripts:

Family Background, Class Background, Motivation/Aspirations, School Education, Choosing University, University Experience and Future Plans. Data relating to these themes is presented below and analysed to identify ways in which family networks, habitus and dispositions may impact on students' academic and vocational pathways.

Table 3.1 Sample Profile: number of participants with stated characteristics

Characteristic	*Number in sample with characteristic*
EU family background	2
LGTB/gender identity issues raised	2
Significant impact health issues raised	3
Self-identified as working or lower middle class	5
Self-identified as middle class or professional	6
Families with parent(s) with university degree	7
Top GCSE/A level grades	9
Strong STEM background amongst family members	8
Chemistry (education/occupation) in previous generations	4
Parent(s) upwardly mobile	10
Firm commitment to science career, academic or industrial	8
Other career possibilities under consideration, including school teaching	3

See also Occupational Table, Appendix

Family Background

Class Origins & Mobility

Three participants self-identified as 'upper working/lower middle class'; two self-identified as 'working/lower middle class'; five self-identified as 'middle class'; one described herself as 'upper middle class' but recognised this status may not translate directly to the British context. These self-descriptions were explained by reference to parental occupations and overall family background. The five participants who described themselves as working class, at least in part, nevertheless cited parental occupations at some remove from the routine and manual employment that might be expected. Ainsley's parents are graduate engineers with British Aerospace Systems (BAE); Anna's father is a mechanical engineer and her mother is a teacher; Gail's parents are self-employed owners of a storage business; George's father is a civil engineer; George's mother is a senior lecturer in HE. This apparent mismatch is consistent, however, with the British Social Attitudes Survey finding that although only 25 per cent of people now work in routine or manual occupations, 60 per cent of Britons regard themselves as working class (Butler 2016). Nearly half of people in managerial and professional occupations identify as working class. Amongst those perceiving their family background as working class, only Gabriella's father (tailor's assistant/chef) is employed in a traditionally working-class job. She describes her mother as an auditor, accountant and business owner.

The six participants identifying themselves and their parents as unambiguously middle class listed professional, administrative or specialised employment requiring degree-level qualifications. As we have seen, Bella's father is a graduate scientist who is a client relationship executive with a global information technology (IT) company. Her mother is also a science graduate who works as an alternative therapist. Carrie's father is director of a landscape gardening company and her mother is a teaching assistant. Daniella's father is a barrister; her mother is an entrepreneur specialising in property investment and has recently acquired a degree. David's parents have lost ground with the collapse of industry in Lithuania but his father was a research metallurgist and his mother is a teacher. Joel's father is an IT consultant for HP; his mother has a PhD and worked for fifteen years as a research scientist at Unilever. Kyle's mother is an environmental health officer and his father is a factory director in pharmaceutical manufacturing.

With the exception of Daniella, whose family has enjoyed high status through at least three generations, there is strong evidence of upward mobility in the family histories documented in this study, with parents achieving status and prosperity beyond the reach of the previous generation, not least because of the different conditions of their lives. As the Occupational Table shows, 9 of the 11 fathers in the sample achieved upward mobility compared with their own parents, with only Gabriella's and Daniella's fathers continuing at roughly the same level. 7 of the 11 mothers in the sample achieved upward mobility in job terms, compared with their parents; 4 others were already in middle class families with professional qualifications and married upwardly mobile men.

There is also a sharp contrast between the work profiles of the mothers and grandmothers in this study, with six of the older women described as housewives employed as routine workers, as secretaries and clerks, or as assistants in bars, kitchens, shops, or in one case, a hair-dressers. One of these was a single parent. Five others were lecturers or teachers, posts sometimes achieved after raising children and working in routine clerical or retail jobs. By contrast, 9 of their daughters are professionals, with responsible and demanding roles. Another is a self-employed business woman; the eleventh is a teaching assistant married to the director of a landscape gardening company. Participants' female and male parents are much more equal in terms of their qualifications and level of employment than those in the grandparent generation on either side of the family. This is significant for family status, income and aspirations. The female contribution to household prosperity, well-being and capital accumulation (of all kinds) is considerable and relevant for social mobility.

Participants are well aware of progress within their families. Ainsley, for example, says that education has enabled the parents to move up in terms of social class; Anna sees her parents' families as 'quite working class' but regards them as having 'done well for themselves, moving on up.' She reports a similar pattern amongst her uncles and aunts. David is confident in his own middle-class identity, but recognises that his father's parents were 'worker class' and different from 'my mum's side' where they were 'more like middle to higher class.'

Gabriella captures the shifting, improvised occupational profile of her migrant family, as well as the underlying theme of owning land and business in Italy, but she also recognises her parents' success: 'I feel like both my parents have come from very little. Both my parents have worked really hard and we have a bit more than they ever had.' Kyle says both sets of his grandparents 'would definitely have been working class but worked up to middle class and my parents have always been middle class.' George's grandparents lived in the suburbs of Birmingham and did 'quite local things…my grandmother didn't do anything aspirational herself.'

Occupational Influences

Our participants' class and occupational backgrounds, though in some respects diverse and varied, are homogeneous in terms of their relative financial security, based on professional or technical expertise, often in STEM-based occupations. The sample includes 8 out of 11 participants with strong STEM influences in their family's employment history. As we have seen, both Ainsley's grandfathers worked as electricians and both parents are Open University graduates with a physics/mathematics background. They 'have worked as engineers forever with Airbus/BAE systems.' Mother is an aerodynamicist with '25 years on plane wings' while father has worked more recently as an IT and science technician. Anna's grandfather and father were both mechanical engineers; three of Daniella's maternal grandparents held chemistry degrees and her paternal grandfather was a university chemistry lecturer. Bella's parents both went to university and have worked in computer consultancy. As she says: 'My parents were intelligent, attended good universities, have science degrees and are more middle class than their original families.'

David's father researched the composition of metals in a Lithuanian university while his maternal grandparents worked in the accounts department of a synthetic silk factory. Joel's maternal grandfather was an electrical engineer and his mother was a research scientist with Unilever for 15 years. His paternal grandfather was a civil engineer and his father is an IT consultant for Hewlett Packard (HP). George says that the grandparents' occupational history 'was nothing really much…not very professional,' although the maternal grandmother eventually worked as a teacher after many jobs that 'were fairly low paid.' George's mother became a nurse, however, and now works in HE coordinating BSc nursing qualifications. Father is a plant manager for an anaerobic digestion facility 'which is quite chemical and quite physical.' Kyle's maternal grandfather was a chemistry teacher and his father works for GSK. He 'has a range of roles in pharmaceutical manufacturing, from factory director to analytical development chemist.' Four of these participants say chemistry is an important strand in their occupational history.

Another three participants come from family networks with less clear links to science-related occupations. Their histories contain, nevertheless, traces of relevant skills and dispositions. Gabriella's 'really big family on my Italian side' has followed diverse migrant occupations over three generations (e.g. chef, production line operative, agricultural labour with olives and grapes) but her maternal grandmother was an accounts worker and her mother is an accountant and business owner. Gabriella says she 'has a mathematical brain and always enjoyed maths.' Gail's maternal grandfather worked as an electrical engineer and her paternal grandmother qualified as a nurse. Her parents are self-employed and run a self-storage business. Carrie's history describes a working-class family from the Welsh valleys, with her paternal grandmother the sole link with a science-related vocation (nursing). The family has a farming background and Carrie's father is a landscape gardener. As she says: 'We're from rural life, it's great.'

These undergraduate histories are consistent with and mainly confirm the findings of the TISME (2013) national survey of 10–14-year olds whose evolving science and career aspirations were tracked over a five-year period.[1] Students from affluent, professional and middle-class backgrounds, with high levels of cultural capital and close relatives in science-based employment, are more likely to choose and continue with science in the sixth form and beyond than their peers. They are also more likely to pursue STEM-related employment. Science capital, defined in terms of a family's stock of science-related qualifications, interests, understanding and social contacts, was found to have a significant but not determining role in leading young people to be interested in STEM and to have a positive view of scientific pathways (TISME 2013). STEM background participants in this sample also reported strongly positive attitudes towards science from an early stage and acknowledged links between family employment and their own orientation. By contrast, the three non-STEM background participants 'discovered' science subjects by enjoying them and doing well at school, in one case, not until the sixth form and A level.

A single occupational or social class focus can be misleading, however. Family role models and dispositions have complex influences and are mediated by individual relationships and interests. In four-fifths of the hundred families Thompson studied, for example, 'there is some intergenerational transmission, and in more than half the transmissions run over three or four generations.' There are broad patterns but nothing

[1] See Chap. 2.

is inevitable. Families offer a choice of models with 'parallel transmission of different occupations' across a network of relatives. Parents may seek to pass on 'something of themselves' but children have to take on what they want from it because transmission depends on two-way relationships (Thompson 1997: 45). Influences can be direct and explicit but also implicit or tacit, with many parents allowing their children space for independent decision. This is important because shifts in circumstances can prompt individuals to make decisions and choices that are based on their own disposition, ability and interests, as much as family history and habitus (Bertaux and Bertaux-Wiame 1997).

Other Family Influences

Although our sample shares many common characteristics, including outstanding academic ability,[2] a passionate love for chemistry and financially secure and upwardly mobile families, there is considerable diversity in family culture and in the values, expectations and strategies adopted by parents. Class differences are significant but the parents' outlook seems also to have been affected by individual histories and their interpretation of past experience. Participants are positive and appreciative of their family life but their stories vary considerably. Ainsley felt that as both parents are STEM graduates it was unavoidable that the family is focused on science. Ainsley felt there was always an assumption 'that I'd go to university.' By contrast, George describes complex interactions and influences in a family that seemed 'weird…just odd.' He takes 'a lot more from my mum than my dad. Especially about acceptance and life lessons. But I get my science side from my dad, definitely but my mum loves the patient side of medicine. I like to think I've got good bits of my parents. And I think my brother and sister do too.'

Anna's parents have regrets about their decisions about HE, with mother wishing her own parents had pushed her more and father accepting an engineering apprenticeship rather than a university place because he didn't think he was good enough. But Anna herself is bright and wanted to go to university from early on, so didn't feel pressure, only the expectation that clever people go as far as they can. Her parents will do 'anything to support my aspirations.' There was pressure to achieve in school in the

[2] This follows almost axiomatically from participants securing entry to a Russell Group university with exacting admission requirements.

sciences and she felt the need to do well in her parents' subjects. She was taught that education was valuable and worth the 'debt and time.' After an unsuccessful year at Cambridge she wanted to leave and was supported by her mother who 'could see how unhappy I was.' Her father did not oppose her departure.

Carrie values the comfort and security of her family and home life. She says there are no constraints, with her parents encouraging her to do 'what I wanted to do' but also reassuring her that 'we're there to catch you if things go wrong.' Carrie 'trusts my parents completely' and believes her family is very protective of one another. She says you were taught to 'stand on your own two feet and know the value of money.' She worked part-time from age 14 and didn't expect her parents 'to pay for me to go to the cinema.' Although Daniella was raised in a scientific, academic family, she experienced a similar level of freedom and support. Far from encouraging her to choose chemistry, her professional chemist grandparents advised against it 'because I won't make any money if I study chemistry.' They thought law or medicine would be a better bet.

Her parents 'were always very supportive of my plans…I can choose what I would like to do.' Her family has a strong belief in 'doing what you like and what you want to do and what you think you will enjoy.' Daniella thinks this is because her mother was forced to study law against her will and didn't finish the degree. 'She didn't want me to be forced to do something I didn't want to do.' David felt pressure from school because he was doing well and there was talk of studying medicine, but says 'for me my parents were completely free about it and if you liked it that's fair enough.'

Joel says his parents' influence 'steered me into STEM subjects.' His father was passionate about computing despite not going to university, and encouraged Joel to 'to get ahead of the game' by exploring HE options. Joel is glad he followed this advice because his first open day at Leicester in year 11 'made me realise computer science was not for me.'

Mothers and fathers are described as encouraging and supportive as well as actively engaged in their children's lives, ready to emphasise the importance of education and to promote positive attitudes, including hard work, self-reliance and being happy. They are also seen as the source of distinctive dispositions, culture and values. These emerged more explicitly as participants discussed their motivation, aspirations and educational decision-making and were said to have been influential in shaping the students' own outlook.

Motivation & Aspirations

When participants reflected on their motivation and aspirations they mainly spoke of personal happiness, job satisfaction and the desire to make a difference in the world, although the need to work for financial security and a comfortable life were freely acknowledged. The desire for high status and wealth was explicitly disavowed, except by David who quoted £100,000 as a benchmark salary for success, and Gail who 'would like to be rich.' But at this stage she is in two minds as she considers the future. She wants to make a difference through an enjoyable, satisfying job, rather than 'be a slave to money.' Gabriella did not refer to status and wealth at all.

Ainsley says the 'biggest motivators' are personal happiness and financial security. Ainsley is 'not interested in being fabulously wealthy' and hopes to make a difference, perhaps by developing a drug 'to help ease someone's pain.' Anna shares this priority but admits being 'motivated slightly by money' because she has not been working hard on a difficult course just to 'earn a pittance.' Her goal is 'to earn more than my parents and give something back, but I realise money isn't everything.' She believes in working hard and doing your best.

Bella cites her decision to leave Cambridge to escape pressure as an example of the importance of personal happiness in her life. She was not prepared to sacrifice happiness for 'classical success' at an elite university, commenting that 'a lot of people thrive in a high-pressure environment while I really hated it. I wouldn't want an intense job.' She is influenced by the ethical, social and environmental concerns embraced by her parents and wants 'to make a difference, that's the big one for me, hence my desire to go into environmental chemistry.' Carrie thinks it is 'superficial to be defined by your status' and believes it is important 'to be happy, satisfied and not wasting my potential.'

Daniella's primary motivation is 'first living in a country I like and feel comfortable in' but like Carrie, she also emphasises her desire for 'a job that satisfies me and is something I enjoy doing,' before reminding herself that it must 'bring me enough money to live comfortably.' Daniella says at school she was determined to forge a career in science and be 'famous and known' but that has become less important for her. She wants to balance well-being and achievement.

Gabriella's hard-working migrant family background seems to have caused her to emphasise work and career rather than dwell on questions of personal happiness. As she comments: 'I know I can be a bit harsh on people that are not as perfectionist as me because I've always had this ethic of working really hard.' Before leaving school, she says she was naïve and 'didn't really think much out of the valley where I lived, you knew everyone.' Gabriella did not expect to leave the community: 'It's a bit of a bubble where I grew up so lots of people leave and stuff.' Now at university, her 'aspirations are far bigger than they were before…I don't want to stay back at home and I've got aspirations to travel and go elsewhere.' She wants a stable job in the future but is keen to find a field to 'stretch the boundaries and make a difference.' She wants to have an impact on the scientific community.

Gail's motivation, by contrast, seems to come from competition. She wants to 'do better than my sister' and to do better than the students at her grammar school who outperformed her at the time. At university she says her 'aspirations seem to have dipped' because she is distracted by living with friends, revises less and works less hard. She feels she has not taken university as seriously as her A levels.

Joel acknowledges that his attitudes come from both sides of his family: 'The world doesn't owe us anything.' Even so, his ambitions are relatively modest. He would like a decent job and wage but 'personal happiness is the important thing.' George says 'it doesn't matter about money if you enjoy doing it.' George is not 'fussed' about income, provided what he does is enjoyable. Kyle's attitude is very similar: 'My parents were more than willing to support me to do what I want to do as far as they can, as long as I'm enjoying myself…I think that would hold for my grandparents as well, they want to know I'm having a good time.'

In reflecting on their youthful vocational ideas, participants explained how initial passions and preferences had to be revised in the light of disappointment, family guidance, school and university experiences and new opportunities. Ainsley wanted to be a pilot and was attracted by the idea of RAF sponsorship but was not eligible because of a personal history of mental health problems and poor sight. Anna was eager to go to dance college but her mother said it would be a waste of her talent. When she achieved an A* in chemistry she realised dance would be a hobby, not her future. Carrie was 'dead set on being a physiotherapist' but hated it on work experience. The work was monotonous and repetitive and she couldn't see herself pursuing it as a career. Her father pointed out that she was always talking about biological chemistry, feedback that influenced her thinking about the future.

David was good at science and believed this would lead to medicine but gradually recognised he was 'more into chemistry.' Gabriella used to dance a lot and wanted to be a ballerina but definitely not a teacher. She was good at all her GCSE subjects, so no clear career path emerged until A level when she decided to become a scientist. As she says: 'So that was a career rather than another just professional job, if that makes any sense?' With no scientific background in her family, Gabriella's ability and enjoyment of science subjects seem to have been a decisive vocational influence. George is different in having a clear science preference from an early age and a sustained interest in becoming a teacher: 'At school I loved chemistry and I wanted to be a chemistry teacher from year 8 and haven't swayed from that. I had a good childhood and I wanted to be a train driver but mum said "you can't see." So that was the end of that.'

Despite a strong family background in chemistry and teaching, Kyle's vocational ideas have been tentative and subject to frequent change, with realism and enticing possibilities in tension with one another. Aged four he met his parents' friends who were nurses and decided he wanted to be a nurse himself. As he reflects: 'Whenever I was in school and somebody came, I thought I wanted to do that too.' At secondary school he wanted to be a programmer and he says that 'I still want to be a programmer; I think that would be interesting.' Now he's thinking about chemistry teaching but is unsure how to go about it because his main interest is in A level work. He has also considered becoming an actor but was afraid that everyone else would 'think I wasn't very good.'

These findings are consistent with our qualitative study of students at two academies in the south of England (Hoskins and Barker 2014). By far the most common and significant aspiration amongst those interviewed is the achievement of personal and professional happiness, by making a difference and enjoying job satisfaction. In this study, as in our previous enquiry, participants' vocational interests and aspirations seem to be woven into their wider family background and identities, and to be heavily influenced by transmitted dispositions, reinforced by examination feedback. Their identities comprise a multiplicity of self-definitions that are subject to variability, discontinuities and turning points as obstacles and opportunities present themselves (Crozier et al. 2008).

As with the British families in Thompson's study, 'it appears that at all social levels family culture, rather than feeding upward mobility, has normally worked more conservatively and protectively. Well-educated parents ensured that their children were educated too' (Thompson 1997: 47).

Our participants' reflections provide many instances of the transmission of cultural capital and persuasive evidence of social reproduction (Bourdieu and Passeron 1977). The parents seem intent on making school work for their families; and wish their children to succeed within acceptable parameters, including university education (Crozier et al. 2008).

School Education

Daniella and David were educated in East European countries before arriving in the UK but encountered no difficulties in aligning themselves with the science curriculum at a Russell Group University. Bella transferred to a Steiner school from a state primary after two years and continued there until age 16, when she moved to a local, independent sixth form. The other eight sample members attended non-selective state schools in England, with the exception of Gail, who moved from a girls' state comprehensive to a grammar where she felt students were 'pushed more' and there was 'more support for going to university.' Only Anna (positive), Bella (positive) and George (hostile) made specific references to their primary education.

From an early stage, participants saw themselves as exceptionally able students with a marked preference for STEM subjects, beginning with mathematics in primary school and then broadening as they progressed through secondary education. Family dispositions towards STEM were soon confirmed by early success and good grades as the examination and test system unfolded. The cohort's academic progress, however, masks their struggles to establish self-confident identities, win the acceptance of their peers and negotiate a path through the social mix and organisational complexities of secondary education.

George 'hated primary school with a passion. I walked in and said I don't want to go. I hated it.' But a state secondary school was 'brilliant for me' because they 'allowed me to be more intellectual and get on with people of the same sort of intellect…they worked a lot towards independence with disability.' George's GCSE and A level grades were 'pretty good' and the choice of chemistry 'wasn't really difficult.' The only thing he faulted was the school's LGBT education 'that was non-existent.' George thinks he would have 'realised my correct gender earlier' with more help but is not alone in retaining unhappy memories of some aspects pre-university education.

Daniella, rated as one of the best 100 students in her home country, describes secondary school as 'the worst period of my life' because she 'wasn't liked by others' and was 'fairly non-standard.' She felt she did not belong:

> I was treated like the odd one out. I've never been very sociable so I was mostly reading books and so on. Also, I was madly frustrated by the teachers. Because I was trying to do my own thing, other teachers were so indifferent; I was asking more complicated questions they didn't want to answer. So that's when I said I wanted to move to Western Europe.

Despite this sense of alienation, Daniella discovered she was good 'in more mathematical subjects' and 'rejected biology because I know it's much more descriptive.' She also had a 'really good chemistry teacher and a really good chemistry tutor to prepare me for the exams.'

Gail was frustrated by her state comprehensive. She was 'bright right from the start' but preferred maths and science to English and Art, subjects that 'sounded like a load of rubbish' to her. But she really liked reading from an early age but it was the story she enjoyed, not the 'taking apart' that happened in English lessons. She felt 'held back by going to a comprehensive' and thought the gifted and talented initiative was 'nonsense' because it was 'supposed to help you go forward but it was rubbish.' Gail was angered by what she perceived as a bias towards less able students:

> For GCSE every lesson had support for C/D background but for A/B stars nothing. They just wanted the pass rate to be right. I didn't feel adopted or encouraged. They offered Media Studies, for stupid people.

Gail was much happier when she moved to a grammar school where she felt there was more academic encouragement and support for going to university. She saw the move as a 'positive shift' for me. Her favourite subject was chemistry, mainly because she enjoyed explaining it to people and 'liked understanding how things worked.' Gail followed A grades at GCSE in all but one subject with top grades in A level biology, chemistry, mathematics and A/S geography. She was disappointed there was no computer science at the school because 'I think I would have liked that.'

Anna is mainly positive about the non-selective secondary school where her mother works. She says she enjoyed it and believes she was fortunate to go there. Her mother exerted considerable influence over this choice, telling her daughter there were only two schools she would be allowed to attend. Anna was thrilled by laboratories and chemicals and this reinforced the idea, formed in year 6, that she has a 'science mind.' She was in the top set for everything and the idea of going to a Russell Group university was 'drummed in at school.'

There was a problem, however, when the school pushed the International Baccalaureate and she was persuaded to opt for a broad-based course by a trusted English teacher. She is glad to have studied chemistry, English and biology at the higher level, and mathematics, psychology and German at standard level. But the effect was to separate her from those who chose A level courses. Her two years were spent with people who would not have been her friends in other circumstances. The result was an unhappy period that affected her grades. Everything 'went sour' and she began to suffer from depression, which runs in the family. Anna experienced a complete breakdown before chemistry. The examination turned out badly and she achieved a lower grade than expected. She only got into Marsden by negotiation.

Bella's mother was 'looking for alternatives' to standard education that would reflect her ethical and environmental concerns. The chosen Steiner school provided 'a more holistic view,' with an emphasis on 'arty and creative things that was really good for me as a scientific person.' She opted to take GCSEs in art, music and drama. But triple science was offered on an extra-curricular basis so she ended up taking additional science instead. The results were 'not as good as they might have been at a different school.' Like Gail, she worried that there was 'less push to help the more able and more help for those who struggle.'

After GCSE, she transferred to a large sixth form college and welcomed the opportunity to have more subject choices and meet more people. She opted for biology, chemistry, mathematics and drama but in the second year dropped drama for further mathematics and began to prefer chemistry because she liked its focus on problem-solving. She was 'interested in how processes worked, chemistry was definitely for me.' Bella was also impressed that the science department at school 'treated us like individuals.' She wanted 'to know everything' and read a history of cancer six or seven times, eventually referencing it in her personal statement.

School education seems to have been much less challenging for the remaining six participants.

David reports direct parallels between GCSE and A levels in Lithuania and the UK. His chemistry teacher at high school was a 'very inspirational person and I got quite a bit of my willingness to do chemistry from her.' He chose science subjects and mathematics 'because I felt that's what I wanted to do after graduation.' The arts subjects did not exert a similar attraction. He encountered few hindrances and 'felt I could do studying fairly easily and nothing was really holding me back.'

Kyle had a long bus journey to the 'closest not terrible school' and encountered early problems with his elder brother's 'enemies who took out their dislike for him on me, initially at least.' He found an ingenious solution by helping the 'enemies' younger siblings with their homework. School was 'not too bad' and he liked the teachers who 'were friendly and they definitely wanted to help you get where you wanted to be.' Kyle's strength was in 'anything with numbers' and 'if it was words, not so much!' This led to A* grades in GCSE mathematics and sciences and a top grade in a free-standing mathematical qualification. Choosing A levels was more difficult. He opted for double mathematics, computer science, chemistry and economics.

Carrie was at a small rural comprehensive, so rural there was a farm with sheep escaping into classrooms. She excelled in science subjects from year 9 and knew that she wanted to pursue a science pathway. Chemistry was the best of her A levels and she talked to teachers about future options. She was very interested in cancer and the science behind biology. Carrie was one of the highest achievers in the school, with excellent GCSE and A level grades. School success and natural aptitude confirmed her interest in science. Her happy and supportive home seems to have compensated for her lack of science background.

Ainsley was a bright student so was able to 'do subjects I was interested in.' Ainsley did triple award science and combined it with history, French and drama. The results (9A*s) were really high and brought pressure to take A levels in double mathematics, chemistry, biology and drama. Ainsley says that 'I fought for drama' but in the end saw the choice as a mistake and dropped the subject after A/S level. Ainsley felt more skilled in science than writing essays and believes this made drama 'the most work' and the 'hardest' subject. The combination of lack of confidence and teacher pressure meant that although 'I knew I'd enjoy it, loved it, I was persuaded further mathematics at A2 would be more useful.' STEM A levels are demanding and hungry for time and reduce the scope for other subjects that might secure a better balance. The guidance and option systems at 16+ and 18+ seem to translate the STEM imperative into a less broad curriculum, especially for the most able students.

Gabriella describes her small rural comprehensive as 'very working class' and 'not elite in any way' but it was good and did well in the league tables. She was comfortable there because 'everyone was kind of roughly around the same wealth as my family.' Her year group had a strong academic background and achieved 'the best scores the school had ever had.' Gabriella herself 'got the second highest grades the school ever had' and 'my friend got the highest grades.' She liked all the subjects, with art, language and history amongst her options. She begrudged dropping others so self-studied astronomy in year 9 and religious studies in year 11. In choosing between arts or sciences 'it was a bit 50/50 but I did really well in chemistry and physics GCSE so I knew I wanted to continue with those.'

Joel reports that his upper school was 'very good.' His mother taught at the school. He was always interested in science and when he moved from middle school 'chemistry cut in.' He 'loved science…the teachers were great and a big influence on me and others.' Joel's parents organised work experience for him from year 10 onwards. He is grateful and thinks it is 'best to start off doing jobs you won't like.' He passed 15 GCSEs, thirteen of them with A* and A grades and opted for A levels in chemistry, mathematics and computing. Joel also reflected on his father and grandfather who both did chemistry degrees: 'I didn't want to do the same as them and then I realised that I did want to do chemistry so I did it.'

These middle-class families, secure in themselves and their children's ability, employ their knowledge and capitals to ensure success over time (Crozier et al. 2008; Whitty 2001). This is achieved not simply by gaming the school and examination system or buying advantage, but also through forms of 'concerted cultivation' that draw on a repertoire of resources and dispositions to shape young people's behaviour and goals (Lareau 2002, 2003). This is why education, working with children's nascent identities and dispositions, as well as with their almost hidden family histories, tends to operate as an agent of social reproduction rather than as a catalyst for improving or transforming society (Mills 2008; Atkinson 2015). Our participants' lives and learning are inescapably influenced by their parents and their family history, with all its twists and turns in particular places and circumstances. Schools do not work as a free-standing variable, to be adjusted to produce the requisite number of people with specified skills and attributes (DfE 2017; Hoskins and Barker 2019).

Choosing University

For many participants, the choice of university followed from choosing chemistry as a degree subject but for others the process of reading prospectuses, visiting institutions and developing course preferences helped crystallise a final decision on where and what to study. Kyle thought about mathematics and chemistry so considered institutions offering a joint course. But when he began to meet academics on open days 'they seemed to suggest that if you want to do maths with something you probably didn't like maths enough to do maths.' Although he was good at mathematics, this advice persuaded him to stick with just chemistry.

Carrie began the Universities and Colleges Admission Service (UCAS) process with biochemistry in mind and talked a lot with her family about decisions. She had the chance to go to Oxford for a week and immerse herself in what biochemistry might be like. She also visited Exeter with her father and began to consider other courses: 'Dad and me went to every open day. We'd talk about what I liked and didn't.' Chemistry emerged as more prominent than biology. Anna's unhappy decision to take natural sciences at Cambridge, together with the unbearable pressure generated by coping with multiple disciplines, enabled her to recognise the merits of Marsden, where there was a strong emphasis on environmental chemistry. Daniella researched UK universities thoroughly, assuming that she would choose to do biotechnology but 'then I thought I actually prefer chemistry.'

A marked preference for a Russell Group university was the start of the UCAS search for many participants. David said the main criterion for him was that a university was strong academically. He perceived the universities in Lithuania to be 'quite poor' and wanted to work in a 'well-established chemistry department.' Gabriella was equally keen to apply to a high-status university, perhaps Oxbridge, but was ill during the whole of her last year and accepted that plummeting grades would not be good enough. She visited a number of Russell Group institutions. Ainsley adopted the same strategy but was disappointed at one place by the unwelcoming and negative attitudes amongst the students who took you round. Gail 'randomly picked' five universities and was a little awed by the number of institutions to think about. She had the vague idea of going to the same northern city institution as her elder sister.

The Marsden open day was decisive for most. Anna says 'Marsden is the only one I fell in love with, I felt it was right for me.' Ainsley found the people around the department 'helpful and enthusiastic' and liked the deputy head of department who was 'lovely and enthusiastic and powerful.' The state-of-the-art equipment was impressive. At Marsden 'it seemed like the people were excited to have new people coming in.' The university was always first choice and this was confirmed by the visit. A 'strong equality and diversity policy' was important in the decision because Ainsley did not wish to be in a workplace that would 'discriminate against me on grounds of being trans.'

As soon as Carrie walked into Marsden she loved it: 'I fell in love in an hour, I knew my first choice straightaway.' She worked hard and was 'over the moon' when she got the required grades. She 'couldn't wait to get to Marsden' and has 'loved it ever since.' She says the university 'now feels like my home, the hub is my family home' but acknowledges that 'it is hard to adjust to not seeing my mum after a bad day.' Gabriella's reaction was similar, despite her illness. When she came to the Marsden open day 'the passion was incredible and it just blew everything else out of the water. I knew this was where I wanted to be. This was my place.' George was equally keen. He came to Marsden with his mother. The deputy head of department 'sold it so well, he was amazing and without a shadow of a doubt this was the best university.' He says he 'just wanted to be around lovely people who just say hello in the morning. That was a big factor.'

Participants were open to new sources of information and guidance during the transition from home and school to the world of HE. Marsden's open days and equal opportunities approach acted as an invitation to join a vibrant science community. But for many, parents continued as a welcome and formative influence on their course and university decisions, and contributed to a pragmatic dialogue about strategies and pathways (Hodkinson et al. 1996). The important role of parents is confirmed by an empirical data set, drawn from a survey of 18–35-year olds[3] participating in the Cultural Pathways to Economic Self-Sufficiency and Entrepreneurship (CUPESSE) project. The findings suggest there is an increased transmission of economic capital today, compared with previous generations. Young people are more reliant on family than in the past, with risks for social mobility for those who depend heavily on family resources for support during their transition from school to work. Parental 'attitudes towards the value of work affect young people's ambitions' (Rainsford et al. 2018: 3). When parents, like Gail's, Carrie's and Joel's, emphasise

[3] The data was collected from a nationally representative quantitative dataset of 3000 young adults aged 18–35.

the link between hard work and money, the CUPESSE data suggests their offspring are more likely to have the 'concrete ambition' to become economically independent (Rainsford et al. 2018).

University Experience

Participants were enthusiastic about winning a place at Marsden and reported positively on their experiences at university. They arrived with a shared enthusiasm for chemistry and a strong but growing science identity, shaped by their many and varied interactions with family members, school systems, teachers and their peers. They differ remarkably, however, in their responses to a new world of hard work in laboratories and lecture halls, challenging and unfamiliar courses, and a rich variety of extra-curricular activities. Anna, for example, found that 'exams and results are the most influential thing.' Course demands mean there is no time 'to do extra-curricular' and her confidence has been 'knocked' by 'struggling a little bit.' As she commented: 'I haven't coped with not being amazing.' By contrast, Daniella worked on two summer projects. One was voluntary in quantum chemistry and 'it was amazing.' She never thought computational chemistry would be so interesting and the experience is influencing her PhD plans. She has also joined lots of societies and is secretary of the clay pigeon club.

Participants were quick to praise the chemistry department. Gabriella says the department 'is very encouraging and they make you feel like a chemist from the start, so you feel involved and part of the chemistry community.' Carrie notices that 'everyone is so enthusiastic' and Daniella has been inspired by one of the professors who found time to teach her about chemistry communication and education. For her that 'opened a new field in front of me and then after my degree I can edit journals or something like it.' Kyle has found the lecturers 'excellent for analytic chemistry' and this has influenced his vocational aspirations. Although Anna has more or less decided that chemistry is not for her it does not feel she's 'wasting my time' because the degree will still be very useful. She has loved the course itself. Ainsley is really enjoying the course and hopes he is on track for a first.

Participants following the four-year Master of Chemistry (M. Chem) degree found the industry placements challenging and enjoyable, as well as opening their minds to new vocational possibilities. Bella has completed an eight-week internship in the atmospheric chemistry lab, funded by a university bursary. She looked at volatile organic compounds with gas chromatography. This has given her a much clearer impression of what it is like to work in a laboratory and has been good preparation for a PhD. Bella is on course for a first. Carrie won a chemistry department bursary to work on admissions and encourage people to come to Marsden. She is keen to be involved but is 'on the fence' about a PhD. Daniella attended a careers symposium with 'one of our alumni who works in laboratory waste management' and is looking forward to an industrial placement next year. The university has shown her 'there's more than just academia or teaching.' Gabriella agrees the industrial placement 'opened my eyes to lots of different things…there's more than I thought there'd be.' David believes the university course has removed his uncertainty about whether to study for a PhD. As he says, 'that's the direction I want to go.'

Joel continues to position himself midway between his parents, poised to transition between chemistry and project management. He hopes to take optional modules in Managerial Economics and/or Management and Industry to prepare him for an entrepreneurial role. He looks forward to the industrial placement and finding out how the chemical industry works. Societies are important for some participants. George runs and enjoys a choir and has found the LGBT group a lifeline. George is also in a brass band and a baking society. These have 'really shaped how I interact with people.' Joel says he is on track for a first but is only 'just scraping it,' partly because of his active life as college tennis captain, a member of Chemsoc and a 'lot of part-time jobs.' He acknowledges that after a 'night out drinking' he is 'shattered and feels that a day is wasted.' Kyle says 'my choir has been the best thing I've done at this university, definitely,' it is 'like an extra family.'

Gail comments on the distractions of living with friends and 'flitting between societies.' She admits revising less and taking university less seriously than A levels. She's busier with 'going out and stuff.' David has adjusted to an increased workload since the first year by 'getting up at 6 and working until evening, just to finish everything.' But despite rising stress levels he is now more confident about chemistry. George was given leave of absence for a year but turned this apparent loss of momentum into an opportunity by accepting an offer to work in a local school as a science technician. The impulse to start work the next day was a good decision because 'I got loads of experience.'

These organised and informal experiences have helped participants deepen and improve their knowledge, skills and understanding of chemistry; and also, to ground themselves in the subject at a high level. They have also prepared students to choose between and follow appropriate academic and vocational pathways. Within the overall framework of a degree course, however, individual experiences vary considerably. Some students are too busy with work to engage with extra-curricular opportunities, while others admit to being distracted by living with friends and going out in the evening. One is finding it hard to cope with not being 'amazing.' But a majority welcomes the challenge of the course and values the stimulus and support of various activities like the choir and the clay pigeon club. Volunteering and bursaries help participants gain insight and unexpected opportunities for leadership, while industrial placements for those on the masters' programme are proving invaluable in providing experience of real-world chemistry.

Future Plans

Academic & Vocational

When asked specifically about their plans, many participants were uncertain about decisions that depend on final examination results and available opportunities a year ahead. Everyone was committed to science careers in HE, industry or teaching but their courses were incomplete and more information or experience was needed to inform the right choice. Some prospective PhD students were unsure about their ability to find and cope with an appropriate research place; intending teachers were slightly more confident about their intentions. Future partners and children were anticipated, with parental models in mind for some.

Ainsley is conflicted about 'what I'm going to do.' The idea of returning to Marsden and working hard to become a researcher or lecturer is attractive but 'I don't know whether I'll be able to achieve that, I'm not sure I'll cope with the stress and environment of a doctorate.' The prospect of 'writing a doctoral thesis by myself' is faintly horrifying. Bella is more confident that she 'will be able to do it' and wants to go into academia, particularly in the field of environmental chemistry. But she understands that PhD study is challenging, 'not an easy 4 years.' The problem, she feels, 'is finding the specific area that excites me.' Bella has used her eight-week placement to increase her understanding of research. Daniella is looking forward to an industrial placement next year. She says 'I'm less determined to become a lecturer' and would consider a very good job in industry. But she also considers that 'becoming a professor would be very rewarding.' Joel wants to retain involvement in university research but envisages transitioning between that and project management.

Carrie is undecided, with research or work for a pharmaceutical company evenly balanced in her mind. She wants to make a difference and improve people's welfare, so although she says a PhD needs thinking about, she 'doesn't want one that doesn't go where I want.' David's thoughts are more straightforward. He plans to do a PhD in organic chemistry. Gabriella is also clear she has a passion for research. But she also havers a little, acknowledging that the upcoming industrial placement could change her mind. She also recognises the importance of finding the right place for further study and securing funding. Kyle chose the four-year master's degree 'to see what research is like' and has discovered that he enjoys it. 'If I do well enough at it then I'll probably apply for a PhD,' he says.

Gail declares she would rather work for charity than a major pharmaceutical company and is considering whether to become a chemistry teacher or a graduate policy adviser with the government. She is slightly resentful that 'the system works well for people who go to private school and have loads of money.' George is more definite, declaring the intention to be a chemistry teacher. Working in a primary school was fun. Although Anna has ceased to be a four-year chemist and has ruled out research, she is confident 'there's a job out there for me, I'm going to careers advice, I love green chemistry.' Participants have a strong chemistry identity and their vocational ideas revolve around jobs where their chemistry knowledge and skills will be used.

Participants' plans for the future are in general consistent their identities as prospective science graduates, with aspirations to fulfil their potential, lead happy lives and make a difference in the wider community by using their rapidly growing specialist expertise in a university or vocational context.

Family

Some participants included partners and children in their future plans. Anna anticipates the challenge of combining children and a career and draws on the example of her mother to assist her thinking. She is worried 'that my career won't really start until I've had children. My mum's career didn't start until she had me and my brother.' Gail expects to get married and have kids and notices that her parents 'share the hours but mother runs the house.' Bella also wants to get married and have children. She has seen a lot of working mums around the department and believes 'you can work around research…other careers are tougher.' Gabriella expects to have her own family in the future, 'a big family because I'm used to that. It would be weird if I didn't have a family because our families are so big and I'm used to everyone being around.' Carrie seems alert to the limitations that come with having children. She says she'll 'need to be with someone who respects my career.' Daniella is 'not exactly a family person.' The idea of a big family 'is not something for me.' She definitely likes being in a relationship but is doubtful about having children. For now, she prefers keeping animals.

Three male participants also discussed family possibilities in their future plans. David's current outlook is similar to Daniella's: 'I would like to have a partner to live with but I'm not certain about having children in the next 15 years. Having children doesn't really appeal to me.' Ainsley 'wanted to find a job that enables me to live with my partner and where we can both work on jobs that we enjoy.' George spoke in similar terms. Success would involve 'living with my long-term partner in a comfortable secure place with both of us working.'

These anticipations of life beyond a first degree do not much assist us in conceptualising women's careers (Powell and Mainiero 1992) and are insufficiently developed to much enhance understanding of the likely impact of raising a family on women and households (Greenfield et al. 2002[4]; Attar 2012). At this stage female participants are beginning to imagine how their careers may be more constrained than anticipated but for the moment they continue to see their futures as a matter for individual responsibility. They hope to postpone families and children until they are properly established. The pervasive influence of family habitus and dispositions is confirmed, however, by participants' references to their mothers and past experience as benchmarks for themselves and their future partners.

[4] See discussion in Chap. 2.

The Genealogical Perspective

This chemistry department sample, at a top university in England, describes itself as 'bright' and coming from encouraging, financially secure backgrounds. They have progressed through school and HE with great success. As the interviews close, they present themselves as more or less fully committed to scientific careers, in HE, industry or teaching. Even Anna, who struggles with not being 'amazing' at university, believes there is a job out there for her, possibly in green chemistry. Gail, who admits treating her degree course less seriously than A level, contemplates teaching chemistry or working in a government department. Otherwise the cohort has postponed decisions until after industrial placements or final examination results, with several participants expressing ambition and doubt about finding a suitable location for further study at doctoral level. They are outstanding young chemists, in most cases the first in their families to attend a leading university. They are poised to provide the specialist STEM expertise and potential leadership the British government is keen to encourage. What are the key influences contributing to their progress and achievement?

Our participants stories include numerous instances of habitus and dispositions operating within families and individuals to shape their choices. They operate as 'broad parameters and boundaries of what is possible or unlikely for a particular group in a stratified world' (Swartz 1997: 103). The participants have varied dispositions that influence their perception of whether a particular social world is viable for them and whether a course of action is possible, plausible or even 'unthinkable' (Swartz 1997: 107).

They belong to financially secure and upwardly mobile, predominantly middle-class families, with high levels of science capital. The parents (9 fathers, 7 mothers) have achieved greater status and relative prosperity than the grandparents, with the women in particular moving from routine to professional or semi-professional roles. Numeracy, technical skills and science-related knowledge and occupations can be traced through the families and clearly impact on participants' identities. Ainsley's parents have moved up; Anna regards hers as having 'done well for themselves.' Bella notices that her parents have science degrees and are 'more middle class' while Joel comments on the professional pedigree in his family. Even Gabriella, with no specific STEM occupation in her background, has a grandmother and mother involved in accounts and accountancy and comments on her own 'mathematical brain.'

Participants reported that their backgrounds were laden with assumptions, expectations and values, even if the pressure they felt was indirect. They recognised themselves as bright and mathematics or science orientated from an early age. Ainsley says the family focus on science is unavoidable; university was always assumed as a destination. Anna experienced pressure from the idea that 'clever people go as far as they can' and felt the need to do well in her parents' subjects. Joel felt that his parents steered him into STEM subjects and notes his father's interventions to promote IT as a career. Daniella's family is steeped in chemistry, but habitus and dispositions, rather than specific advice to avoid chemistry because it is not especially well played, seem to have influenced her thinking. Mothers and fathers were said to be actively engaged with their children, promoting the value of education and encouraging hard work, self-reliance and being happy.

Participants stressed happiness and job satisfaction rather than wealth or status as personal goals, with family habitus fostering the idea of high-level intellectual challenge, scientific excellence and professional status. Carrie wants to be 'happy, satisfied and not wasting my potential' and to be 'challenged in my work.' Gabriella would like to work in a field where she can 'stretch the boundaries' and 'make a difference.' Families may seem conservative and protective but they are also concerned with professional achievements. Parents intervene and support at every stage to ensure the best possible outcomes. Bella's mother sought out alternative education; Carrie's Dad travelled to university open days with her. Joel hopes to follow his father in juggling entrepreneurship with high level expertise. Reflecting on their future plans, participants are heavily influenced by family models. Gabriella wants to have a large family because that is what she is used to; Anna worries about her own career and the fact that her mother didn't really get going until after children.

The participants' familiarity and confidence with technical, scientific subjects at school, together with their success in tests and examinations, helped the process of building science capital in the next generation and reinforced their bias towards chemistry. Teachers are mentioned, occasionally as a source of inspiration and encouragement, but overall the data gives an impression of gifted young people inevitably progressing towards top GCSE and A level grades, with school options and priorities guiding them to elite science and elite universities rather than opening their minds to new possibilities or developing other passions. Sometimes school proved more hindrance than help, with George hating primary and

Daniella feeling she was 'treated like the odd one out.' Gail was frustrated and bitter about her comprehensive; Anna experienced social dislocation when she was persuaded to opt for the International Baccalaureate.

The intensive, demanding chemistry course at Marsden is reported as well-taught, exciting and challenging, with industrial placements particularly valuable for introducing a wider range of vocational possibilities. Even amongst our cohort of able undergraduates, however, there are one or two who have begun to recognise how hard it is to cope with work at this level while committed chemists find there is less and less time for anything but work.

This study shows that the influence of family networks is considerable and long-lasting. Family habitus, capital and dispositions condition and sometimes shape young people's responses to school, teachers and the curriculum, not in a predetermined way but as active ingredients in a complex equation. This genealogical perspective challenges the belief that individuals are isolated agents, armed only with a personal set of unique talents and the capacity for hard work. Instead, participants are revealed as belonging to family networks where small and sometimes large advantages are accumulated over time, enabling members to manoeuvre towards new opportunities when conditions permit (Gladwell 2008). Some families are able to build on the advantages they have already acquired and move forward in successive generations, like the scientists in this study, whilst the less fortunate remain mired in hardship that becomes more rather than less intractable (Mare 2011). School and university are vital enablers for young scientists but work with families that provide the preconditions and continuing influence necessary for success. Middle-class students from a technical or scientific background have a precious, active endowment that enables them to thrive and exploit the educational opportunities they encounter (Archer et al. 2012). For the upwardly mobile participants in this study, gender identity seems to have had little impact on their ability to draw on family science capital and take full advantage of science education at school and college.

The evidence gathered from the case study sample confirms that propositions (1), (2) and (3) are valid and capture the main influences on participants' decisions and choices as well as helping to explain their chosen academic and vocational pathways. Underlying social structures (inequality, family circumstances and dispositions, social class) constrain the scope for individual agency and condition life chances without determining them. This is why the issue of equality, discussed in the next chapter, is so important.

References

Archer, L., DeWitt, J., Osborne, J., Dillon, J., Willis, B., & Wong, B. (2012). Science Aspirations, Capital, and Family Habitus: How Families Shape Children's Engagement and Identification with Science. *American Educational Research Journal, 49*(5), 881–908.

Atkinson, A. (2015). *Inequality. What Can Be Done?* Cambridge, MA: Harvard University Press.

Attar, N. (2012). *Susan Greenfield on Women in Science: Not Much Has Improved in 10 Years.* BMC/Springer, Blog Network. Retrieved September 1, 2018, from https://blogs.biomedcentral.com/on-biology/2012/03/08/susan-greenfield-on-women-in-science-not-much-has-improved-in-10-years/.

Bertaux, D., & Bertaux-Wiame, I. (1997). Heritage and Its Lineage: A Case History of Transmission and Social Mobility Over Five Generations. In D. Bertaux & P. Thompson (Eds.), *Pathways to Social Class: A Qualitative Approach to Social Mobility* (pp. 63–97). Oxford: Clarendon Press.

Bourdieu, P., & Passeron, J.-C. (1977). *Reproduction in Education, Society and Culture.* London: Sage.

Butler, P. (2016, June 29). Most Britons Regard Themselves as Working Class, Survey Finds. *The Guardian.* Retrieved April 24, 2018, from https://www.theguardian.com/society/2016/jun/29/most-brits-regard-themselves-as-working-class-survey-finds.

Crozier, G., Reay, D., James, D., Jamieson, F., Beedell, P., Hollingworth, S., & Williams, K. (2008). White Middle-Class Parents, Identities, Educational Choice and the Urban Comprehensive School: Dilemmas, Ambivalence and Moral Ambiguity. *British Journal of Sociology of Education, 29*(2), 61–272.

DfE. (2017). *Unlocking Talent, Fulfilling Potential: A Plan for Improving Social Mobility Through Education,* CM 9541.

Gladwell, M. (2008). *Outliers: The Story of Success.* New York: Little Brown.

Greenfield, S., Peters, J., Lane, N., Rees, T., & Samuels, G. (2002). *Set Fair, A Report on Women in Science, Engineering, and Technology.* London: Department of Trade and Industry.

Hodkinson, P., Sparkes, A., & Hodkinson, H. (1996). *Triumphs and Tears: Young People, Markets and the Transition from School to Work.* London: David Fulton Publishers.

Hoskins, K., & Barker, B. (2014). *Education and Social Mobility: Dreams of Success.* London: Institute of Education Press.

Hoskins, K., & Barker, B. (2019). Social Mobility: The Potential of a Genealogical Approach. *British Educational Research Journal, 45*(2), 238–253.

Lareau, A. (2002). Invisible Inequality: Social Class and Childrearing in Black Families and White Families. *American Sociological Review, 67*(5), 747–776.

Lareau, A. (2003). *Unequal Childhoods: Class, Race, and Family Life*. Berkeley: University of California Press.

Mare, R. D. (2011). A Multigenerational View of Inequality. *Demography, 48*(1), 1–23.

Mills, C. (2008). Reproduction and Transformation of Inequalities in Schooling: The Transformative Potential of the Theoretical Constructs of Bourdieu. *British Journal of Sociology of Education, 29*(1), 79–89.

Powell, G., & Mainiero, L. (1992). Cross-Currents in the River of Time: Conceptualizing the Complexities of Women's Careers. *Journal of Management, 18*(2), 215–237.

Rainsford, E., Rawlings, L., & Mistry, L. (2018). *The Role of Family in Social Mobility: Impact of Family Resources on Young People's Progression*. Youth Employment UK. Retrieved July 7, 2019, from https://www.youthemployment.org.uk/dev/wp-content/uploads/2018/03/family-Social-Mobility-Final3.pdf.

Swartz, B. (1997). *Culture and Power: The Sociology of Pierre Bourdieu*. Chicago, IL: University of Chicago Press.

Targeted Initiative on Science and Mathematics Education (TISME). (2013). What Influences Participation in Science and Mathematics? Retrieved August 28, 2015, from https://www.kcl.ac.uk/sspp/departments/education/research/aspires/TISME-briefing-paper-March-2013.pdf.

Thompson, P. (1997). Women, Men, and Transgenerational Family Influences in Social Mobility. In D. Bertaux & P. Thompson (Eds.), *Pathways to Social Class: A Qualitative Approach to Social Mobility* (pp. 32–61). Oxford: Clarendon Press.

Whitty, G. (2001). Education, Social Class and Social Exclusion. *Journal of Education Policy, 16*(4), 287–295.

CHAPTER 4

Equality Policies and Initiatives at Marsden

Abstract In this chapter, we draw on staff interview data to examine the chemistry department's equality policies for staff and students. First, there is acknowledgement of the equality progress made in the case study department prior to gaining the Athena Swan Charter (ASC). Second, there is discussion of the institutional policies implemented as a consequence of the ASC, including shared parental maternity/paternity leave and unconscious bias interview training. Third, we examine the department's widening participation policy, which aims to improve non-traditional student representation in the department's undergraduate programme. Fourth, there is an examination of the equality initiatives aimed at supporting staff and students, for example, female representation across all areas of work including external speakers, external examiners and student role models.

Keywords Role models • Widening participation • Non-traditional student • Equality

K. Hoskins, B. Barker, *STEM, Social Mobility and Equality*,
https://doi.org/10.1007/978-3-030-49216-8_4

Introduction

There have been numerous developments since the 1960s that have attempted to improve equal opportunities, including the ASC and widening participation policies, which aim to diversify the student and staff body in HE in England. But to what extent have these policies and initiatives opened up the study of chemistry to non-traditional[1] students? To what extent is proposition three, identified in Chap. 2, correct; do gender and ethnicity have a major influence on self-identity and aspirations? Is proposition four accurate; is identity a stronger influence than equal opportunities policy and its effects, and does this explain why the pattern documented by Smith and White (2011) persists despite much apparently worthy policy activity?

In this chapter we draw on staff interview data to examine Marsden's chemistry department's equality policies for staff and students before and after they achieved the ASC. We begin the chapter by acknowledging the equality progress made in the case study department prior to gaining the ASC and then examine the resultant institutional policies implemented, including shared paternity leave, maternity leave, flexible working and unconscious bias interview training. Next there is an examination of the staff representation equality initiatives, which focuses on female representation across all areas of work including external speakers, external examiners and student role models. Finally, the chapter discusses the department's widening participation policy, which aims to improve non-traditional student representation in the department's undergraduate programme and is also a key aim of the ASC (Table 4.1).

Equality at Marsden

All of the staff acknowledged that gender equality policies were in place at Marsden prior to the award of ASC. The staff touched on areas where they felt supported, including increased gender equity in the recruitment of senior staff, supportive family policies (in relation to children and other caring commitments) and a democratic, non-hierarchical working culture. Taken together, these supportive practices created an ethos where staff

[1] We use the term non-traditional students to encompass school leaver, mature students, and other historically excluded groups, such as women, the working classes and those from minority ethnic backgrounds (Crozier et al. 2008).

Table 4.1 Staff job title and joining details

Pseudonym	*Job title*	*Date joined*
Audrey	Schools Outreach Officer	Not known
Ellen	Research Laboratory Assistant	2003
Hayley	Department Administrator	2002
Linda	Employability and Diversity Officer	2012
Noah	Lecturer	2000
Rachel	Research Fellow	2008
Robin	Professor	1999
Susie	Lecturer	1997

perceived they were enabled to work, regardless of their identity and their commitments outside of work.

In terms of the gender representation, Ellen explained that when she had joined the department in 2003, she became aware of the relatively strong support already in the department for female chemists. She explained that 'there were two female professors, which, compared to what friends had in [...] other departments, they thought it was quite unusual.' For this reason, Ellen suggested the department had perhaps 'always been a bit sort of ahead in that regard.' Similarly, Robin, who had joined Marsden in 1999, had also noticed that gender representation amongst the academic staff was better than in other institutions he knew 'I wouldn't say there were huge numbers of female academics. There were more than at some departments. Not as many as we have now.' Whilst there was still some way to go to achieve gender parity in the academic staff body, Marsden was perhaps further ahead than other chemistry departments in English universities at that time. Robin told us that whilst 'it certainly wasn't that we had unusually large numbers of female academics in that period,' he also acknowledged 'we probably had more than some traditional departments would have, which would often be zero back then ... We probably had four or five women in academic positions at the time, which was probably a bit unusual.' According to Bagilhole's (2007) research, women in natural science disciplines in the early 2000s were likely to be in a gender minority of 2–6 per cent and little has changed since (Smith and White 2018). Given the context of unequal gender representation amongst academic staff, the department was perceived by participants, several of whom had worked and studied there in the early 2000s, as an inclusive and enabling environment.

Not only was the ethos perceived to be supportive by all the staff, but there were some policies in place to support women before the ASC was implemented, for example, the maternity leave policy. Susie, who has worked in the department since joining as a post-doctoral researcher in 1997, experienced an environment and ethos that supported women academics well before discussions about the ASC. She described 'an open ethos in the department from quite a long time ago. […] I think just wanting to support the women in the department.' When Susie had her first child in 2000, a small number of women academics at different and more senior grades to her had already benefitted from the maternity leave policy. Reflecting back on when she joined Marsden, Susie recalled there were 'these very kind of early things I think that certainly the department didn't have to do,' but they did do because there was 'an awareness from the senior staff, you know, I'm putting words in their mouths, I don't entirely know where it came from … but probably just to try to put into place some ways of supporting those women in a positive sense.' Susie felt the maternity policy was due to senior colleagues acknowledging that women needed support to take time out from their career to have a child.

Hayley also said there were several support structures in place, and that these helped colleagues with caring commitments. She explained that 'we'd already got a lot of initiatives. And I would say that the culture here was quite strong in that area anyway.' Many staff children attended the on-campus university nursery and this contributed towards their perception that Marsden is family friendly. Audrey acknowledged the family friendly atmosphere and noted that she had seen how this extended to male and female staff and to the department's students:

> What I really like about the department here, is that it does have a very family friendly policy, and it's not just for women, it's for everybody. And that, you know, little things, like changing the times of lectures that they were encouraging the whole department to come to, so that they finished before the time when people had to pick their children up from the university nursery. Because, otherwise, you've got this great clatter about 15 minutes before the lecture finished, as people had to nip out to go and pick their kids up.

These apparently small changes seemed to create a strong impression on the staff we interviewed as they all spoke positively about the enabling structures in place to accommodate childcare commitments. Staff benefitted from the family friendly atmosphere and we noted several examples where staff felt that their caring commitments were enabled rather than frowned upon by their colleagues and senior leaders.

The gender support acknowledged by these participants was in place well before any discussions about ASC began around the year 2000. This suggests the environment was somewhat exceptional in comparison to other chemistry departments in English universities at that time. Several staff had benefitted from gender equity policies, particularly in relation to their caring commitments for either children or elderly family members. Greenfield et al.'s (2002) report on women in STEM found that women's progress was hindered by traditional family roles, responsibilities and expectations, but according to a number of participants Marsden was well ahead of its time in supporting and enabling women's careers. The department acknowledged and supported women carrying out a disproportionate share of caring duties.

There was also an inclusive working ethos, described by staff as permeating the department. Robin recognised that since he started working at Marsden there had been, and continued to be, an inclusive work environment. Robin described the institutional structures as enabling, non-hierarchical and democratic:

> Within the structure of the department at that time, it was a relatively non-hierarchical structure. The whole university was quite non-hierarchical compared to what I was used to. There were no faculties, there was the Vice Chancellor, you know, senior management, then there was the head of the department, and there was very little sub-structure underneath the head of department then either. It was a fairly flat structure underneath. So, it felt a fairly democratic kind of place where lots of different people got listened to.

The flat structure provided the space for people to 'listen to one another, sort of respecting one another's opinions and so on' (Robin). Robin discussed how junior and senior staff alike were encouraged and supported to make contributions to debates about the workings of the department in staff meetings.

Noah had started at Marsden as an undergraduate and progressed through his doctoral and postdoctoral studies to a lecturer post. He similarly noticed a strongly inclusive ethos in the department throughout his time and described the existing ethos as matching the ethos and aims of ASC. He perceived that 'as soon as the scheme was begun, there would have been people in the department who would have realised that the scheme meshed very well with ... what they would have regarded as Marsden's ethos.' This meshing together was possible according to Noah because 'the scheme matches the ethos' in the department. The alignment and synergy between the existing culture and ethos in the department and the aims of the ASC can be seen in Noah's perceptions of the institutional ethos. Indeed, a key strength of the ASC lies in its ability to drive conversations about gender equality at all levels across institutions and its ability to drive positive changes in culture and ethos (Buckingham 2019).

The data confirms that prior to the case study chemistry department applying for Athena Swan recognition, there was a range of equality work already taking place. In the perceptions of our participants, Marsden was ahead of other institutions in terms of providing supportive structures and policies for the facilitation of gender equity, family responsibilities and a work life balance.

Policy Changes and New Initiatives Developed Through the ASC

New institutional policies were introduced and implemented to comply with Charter principles. New policies included, for example, shared maternity/paternity leave and unconscious bias interview training. The decision to apply for recognition was taken by the then male head of department who identified it as a key initiative for the department at that time. He began discussions with staff about the ASC and application process in 2000, leading to Athena Swan recognition in 2007. Reflecting the application process, Robin explained that

> I perceived [Marsden] as being quite different to either of those [other universities] in terms of on the ground sort of policies and procedures. So, I think we're starting from a position of relative strength when looking at those equal opportunities issues actually, compared to some departments.

The ethos and culture at Marsden may have made the process of obtaining an award easier than at other institutions where Robin had experienced 'very traditional chemistry departments,' with limited staff diversity, particularly in relation to social class background, gender and ethnicity. Indeed, this lack of social class and ethnic diversity was an issue and something that several staff commented on. One aim in applying for an award was to further diversify the student and staff body.

The award process involved building on the department's inclusive culture and environment as well as considering additional policies and initiatives to meet the criteria. Participant perceptions of the impact of these policy changes on the working environment, ethos and culture at Marsden are discussed below.

Shared Parental Leave Policy

Part of engaging with the Athena Swan process involved the introduction of a shared parental leave policy, which aimed to redress the widely acknowledged gender inequity that exists for female academics as they attempt to balance their work and home lives (Hoskins 2012; Smith 2011). The aim of the shared parental policy was to ensure that leave could be equitably used by both parents. Noah commented that when his two young children were born, he benefitted from having extended parental leave and commented that 'compared with, you know, friends of mine who work in the private sector, it is very generous and I'm grateful for that.' Noah explained that he has two children and in between his two spells of paternity leave the rules in the department had changed and he was not entitled to the same amount of time for the birth of his second child. But the head of department called him in and explained that, whilst he was a week late for the previous paternity allowance, they would extend it to him anyway.

This generous gesture was, according to Noah, a reflection of the 'ethos that permeates the department, that maybe we don't have to do this, but why not, you know, why shouldn't we and take that extra step in favour of staff in these matters.' Recalling this period of time, Noah acknowledged that whilst these were relatively minor benefits for him in the scheme of things 'it certainly made the difference to me at the time.' Rachel said the major benefit of the ASC for her followed the birth of her two children. She explained that her 'whole perception changed when I fell pregnant and was going to start a family because I had to start thinking about how things were going to work.' When faced with the birth of her first child she recognised 'that the department did have a strong equality and diversity policy.' When Robin and his husband recently adopted their first child he reflected on the ease he experienced in taking his parental leave and how beneficial this was for his home life.

Incremental adjustments like these were an important benefit for staff at a relatively low cost for the institution. They are changes that have nevertheless made an enormous difference for individual staff members at times when they needed additional support. These incremental, small allowances seemed to accumulate in staff perceptions and contributed towards their sense of working for an inclusive and supportive employer.

Maternity Leave Policy for Doctoral Students

One of the highest drop-out rates in the department occurred between the completion of PhD studies and gaining a post-doctoral researcher post. This was most marked amongst female students. The limited provision for parental leave was also an issue for colleagues, a concern also reported in recent evaluations of the ASC (Buckingham 2019). The department decided, therefore, to take parental leave one step further than demanded by Athena Swan and implemented a maternity leave policy for doctoral students and postdoctoral researchers, following the pregnancy of one of the department's PhD students. Susie explained that this prompted the department to put a maternity policy in place for PhD students. It was 'one of these things that, you know, until you kind of come across it … and so, there is no provision.' Susie said this was a key issue for the grant funding bodies to consider as they make no provision to cover maternity leave for PhD students. However, because of available institutional funding the department was able 'to find the money to cover, so we were able then to provide that student with some … a kind of stipend through the period when she was actually on leave' (Susie). This led the department to develop a policy for PhD students that provided some financial support for maternity leave should the need arise in future. The policy was a good example of a gender specific intervention aimed at improving the participation of female students working at and beyond doctoral level.

Flexible Working Initiative

Perhaps one of the most significant initiatives to be introduced through the ASC process was the flexible working initiative, which applies to staff from post-doctoral researcher upwards. Robin talked at some length about the significant benefits that this initiative had created for all staff to enable them to deal with a variety of caring commitments. He explained that

> In terms of policies, at the staff level, and it affects post-docs as well, the single, in my opinion, most important policy has been flexible working around childcare and part-time working around childcare has been the most beneficial. When I go and talk to people at other places, they find it very surprising that within the department we have six or seven people, academics who are all part-time working because they have different care responsibilities, elderly parents or children, or whatever. They might be male, they might be female. The majority are female and have been female. But I've taken that leave when we adopted and there's another guy in the department who's doing it at the moment while his kids are young, because he's doing the primary care role.

Whilst Robin suggests that some males have benefitted from this initiative change, he notes that predominately it is women who have been able to work around their caring commitments and so maintain successful careers. This initiative, we suggest, has implications for rebalancing the socially constructed gender responsibilities typically taken up by women working in the department. Burr (2015: 128) suggests that it is the 'repeated performance of gender that gives it its sense of solidity and naturalness,' and that only by disrupting normative gender enactments, in this instance the distribution of caring commitments, can the social construction of gender can be reconstructed differently to open up more inclusive and equal spaces and opportunities for women at work.

Maybe what makes the initiative so unique and successful is the ease with which staff were able to apply for, and be given, these flexible working arrangements. Robin explained that it was a very simple process whereby 'if you ask for part-time working, you will get it. If you ask to stop part-time working and go full-time again, it's guaranteed really.' Staff across all grades were required to give only a few weeks' notice to either increase or decrease their working hours. As Robin noted, this initiative tends not to work in many institutions:

> Most places that I go to talk to people in chemistry, at least they might have a staff member who's been part-time working and had a horrible experience, and no one else ever wants to do it ever again. And so the fact we have six or seven and there's almost a rotation of people doing it and not doing it, you know, I think many people almost think it's the thing that you would do if you had a young family. And the department would be a supporter of that. That creates a real ethos, which is nice.

What was striking in colleagues' reflections on this initiative was their recognition that the initiative works really well because there are structures (including the budget) to operate the policy so that no staff member has been refused or delayed in changing their working patterns. It is worth noting that the department is quite a large one and has resources that can be redistributed to allow for flexible working. Staff perceived that, once established, an inclusive and supportive ethos that permeates through the department is something that can be improved upon and developed relatively easily through initiatives such as flexible working.

Rachel noted that when she became pregnant with her first child she 'never worried once that I would struggle to come back to work, I wouldn't be able to reduce my hours, I felt very confident that precedents had been set, it was family friendly so I never worried.' Given the pressure female academics can experience with the birth of their first child, this policy was significant in reducing potential stress and anxiety about going back to work and how they would manage the balancing act of work and home life demands (Hoskins 2012). As Rachel noted, she never worried about asking if she could return to work part time, or if she needed flexible days during the whole of her maternity leave. She was confident in the knowledge that any flexibility would be accommodated by the department.

Noah also reflected on the department's ability to accommodate some colleagues: 'slightly strange arrangements of hours for certain periods of time—some people have gone, for instance, to eighty per cent or something like that.' Whilst some people's workloads have needed to be 'reconfigured in order to allow for this,' a process that has not always worked perfectly, the department 'has enough people that you can kind of shuffle a few things around and absorb that in order to make the arrangements.' Reflecting on his time in the department he personally knew of five colleagues who had benefitted from the flexible working initiative.

Linda explained that they are not allowed to refer to flexible working as a policy, but rather, it should be referred to as the 'Part Time Working Assurance,' to take account of some of the logistical challenges that it can cause from time to time amongst the staff body.

> It's not a policy and we're not allowed to call it a policy. So, if a person is on, say, a full-time contract, and wanted to drop their hours, for example, for reasons of childcare, there would be the expectation that, subject to all the relevant finance and everything, they would be able to put their hours back up in the future, that that money would be budget protected. That's

something that we do as a department, obviously, there's some logistical issues with that, but that's the way that we try to do it and that's why we're not allowed to call it a policy.

Linda reflected on the potential logistical issues sometime associated with enacting the flexible working initiative, due to movement in the staff body because of secondment, sabbaticals and long-term sick leave. This unpredictability occasionally created difficulties for staff in picking up elements of each other's workloads.

Susie also talked about the department being a beacon in implementing flexible working before the university adopted it as a working practice more widely. She described the flexible working initiative as a 'guarantee for all staff. So, any staff member that wants to work part-time, can work part-time.' The scheme works because they are always able to change their hours back up to full time whenever they want to. Susie explained that if a member of staff goes down to working 'fifty percent … it's really a formality that you just come in and request to go back up to sixty percent or seventy percent, or eighty percent,' and she Susie pointed out that staff are not going to be told 'OK, you've moved on to fifty percent, you have to stay on that now.' The flexibility to move from a fractional post back to full time and then back down to a fractional post was mentioned as a beacon of good practice at Marsden. According to our participants, several chemistry departments across England recognised the excellent equality work in the chemistry department at Marsden.

Hayley benefitted from flexible working when her children were young and she moved to part-time working. She then changed from administering a research group to becoming the departmental administrator. Her workload then increased significantly and her new head in the chemistry department commented, 'I know you are part time, but I know that you do full-time hours.' He noticed that she was coming in at weekends to get through all of her work, a situation that he was uncomfortable with. Hayley reported his comment: 'It doesn't seem fair to me that we're paying you a part-time salary …' So he advised that he would pay her a full-time salary but 'you have the flexibility if you want to take an afternoon off to go … football match, sports day, whatever it is. You know, obviously not every week! … But as and when you need it.'

This degree of flexibility was invaluable to Hayley and she talked at some length about the real difference this made to her work-life balance, not just in practical terms but in terms of how she felt about her job. She said it was the first time in her working life that 'somebody had sort of put me at the centre … and my needs, as opposed to before I was the one that was made to feel like I was inconveniencing everybody else by asking for these things and then just being told categorically no.' This recognition from her head of department that she was going above and beyond her contractual hours to get the job done had 'a really big impact' and Hayley was then able to use her positive experiences of flexible working to advocate that this initiative be extended within the university.

As department administrator, Hayley was often the first point of contact for students, and her positivity towards her place of work was something she could communicate to the students. She told us that

> For students coming in, you know, whatever their background, they should feel they're coming into a friendly, welcoming department that will adapt to them as opposed to them having to adapt to a very rigid set of rules and regulations within the department.

Promoting this sort of institutional ethos focused on inclusivity is at the heart of what the ASC aims to initiate and implement in universities. The example of Hayley's experience and her attempts to communicate this warmth and flexibility to students, from undergraduates to post graduates, fits well with the ASC's aim to address gender inequalities through commitment and action from everyone at all levels of the organisation. It is an important contribution to tackling the unequal representation of women in science by changing working cultures and attitudes (AdvanceHE 2019).

Staff Appointment Practices

The careful efforts made to increase the number of appointments from under-represented groups is a prominent example of the department's determination to change culture and attitudes. Hayley commented that in recent times: 'we've done a lot of work on unconscious bias.' The department has invested time and energy into developing unconscious bias training to ensure that staff responsible for short-listing and interviewing candidates are aware of their bias and able to mitigate against it through their increased reflexivity and peer observation.

Linda led the initiative and prior to our interview she was involved in training colleagues to be aware of the possible influence that unconscious bias can have in the recruitment process. She told us that:

> We're just doing some unconscious bias observing of recruitment, which we're trying out. That's really helped me in two ways, one, because it's really interesting to observe. But also, from an employability point of view, actually sitting in and watching shortlisting, and watching candidates being interviewed, allows me to give really good quality advice to the people that I advise. Because, you can see what's going wrong, and you can see things done to address any issues.

Through the training the staff were able to reflect on and address issues of unconscious bias in relation to recruitment of new staff in an effort to diversify the staff body. Over time, the aim is to continue diversifying the staff body to include a greater number of people from minority and working class backgrounds. Staff felt that the student body could benefit from increased diversity.

Focusing on Representation

A further strand of the department's strategy was a determination to ensure a genuine mix of backgrounds amongst guest speakers invited to talk in the department. Staff said there was commitment to ensuring different groups were well-represented to ensure a wide range of academic role models for the next generation of scientists to aspire to. Linda gave the example of a recent series of career talks for undergraduate and postgraduate students, where speakers have been a 'real mixture of people coming, both internally and externally, to come and speak at that. And, we've been quite open about the fact that that's about diversity, we've had LGBT members of staff. Our last speaker was a female Mexican, Catholic, Lesbian engineer, who was a Professor.'

Robin perceived that at undergraduate level it was easier to ensure representation by making sure that 'invited speakers are representative female/male, other minorities as well.' He suggested another difficult area that requires continuing and careful attention:

> And external examiners for PhD students ... because, you know, scientists work in teams, so I have ten working for me. If they see that every external examiner that comes in to examine someone in my team is male, what message does that send to my team, which is actually a forty percent female team at the moment? It sort of suggests that the people who sit in judgement over their careers in the future will all be men.

He noted that staff have to think 'really carefully about those kinds of things. So the sort of representational things have been very important.' Through carefully targeting a wide range of national and international speakers and external examiners from varied backgrounds and with diverse experiences, the normative social construction of a scientist can be challenged (Burr 2003).

Staff appointment panels were identified as an area where more attention to gender equity was required. As Robin noted, an important shift has been on 'changing what we value at recruitment panels, and we've worked quite hard on that as a department.' These changes have included unconscious bias training as discussed above, but also a different approach to shortlisting candidates. Robin explained that now, when looking at the shortlist, they ask 'does this fairly represent the people that were applying for this job? Or is there a whole grouping that have disappeared as we've gone through the shortlisting? And if they did disappear, why did they disappear?' If the panels note that particular groups have disappeared in the process they go back through the paperwork and 'look at the very best ethnic minority candidate, because we've got none on the shortlist. Or the very best female candidate, just to check that we've not overlooked aspects of their CV that were good.'

Linda noted the initiatives aimed at critically reflecting on the 'makeup of recruitment panels, for example. So, trying to make sure that there's a female member of staff on interview panels.' These attempts were galvanised because of an incident the department encountered over post-doctoral appointments. Linda explained that at another institution that jointly recruits students with Marsden 'there was an all-male panel, and that was very swiftly sorted out for the next year, because we weren't pleased. I think, now it has become unacceptable in the culture of the department, and we're certainly conscious that it looks bad.' The concern around this was related to the perceptions of existing students but also the department's ability to recruit female students. Linda noted that if the department had a 'panel of four male academics interviewing a student, that can be very intimidating. So, we are conscious of that, and the Chair of the Graduate School and myself, we've been working on these kinds of things.' These efforts seem to be paying off because the department has gradually improved its representation of female professors to around 25 per cent.

These initiatives to improve representation in terms of class, gender and ethnicity led Susie to comment that it is 'very much a department that feels like a community' and an inclusive and enabling community. These initiatives also resonate with government policy attempts to widen participation across post-16 and HE to diversify those young people choosing STEM subjects (Smith 2011).

Fixing the 'Leaky' Pipeline?

Staff offered a variety of reasons for the loss of well-qualified women from academic careers in chemistry, usually at the post-doctoral stage. These issues include the dominant social construction of scientists, which remains male and the high drop-out rate that occurs after the successful completion of a doctorate.

Several staff still identify gender as a key issue deterring some females from considering an academic career in chemistry. Hayley feels that the 'perceptions that males make better scientists, in many cases are still there, and they're very ingrained. And it's almost part of the culture, the wider culture that we live in.' She thinks there is a dominant view that 'the males will be the scientists with the great ideas and that women will perhaps do the more caring roles, the nurses and the teachers.' This perception was echoed by Ellen who similarly noted 'there is a big problem with the general association of, if you say to someone, "what does a scientist look like?", they generally always picture it as being a male.' The social construction of the male scientist remains powerful and the lack of female role models can be off-putting to potential new female recruits into chemistry, a finding that was evident in Le-May Sheffield's (2004) work.

Participants believed that continued improvements in gender equity depend on initiatives that help students progress from graduate to postgraduate study. The majority of female students tend to drop out of chemistry once they have achieved their undergraduate degrees, according to the staff we interviewed. Robin explained that:

> One of the things we did is, I think we identified as a department where the big issues lay in progression, particularly of women as a group along the scientific pathway. And the biggest barriers actually are between PhD and post-doc. So I think a lot of focus of the department has been at the PhD, post-doc interface. How do we work with PhD students to change their perception of what an academic career might involve or might be? So I think most of the work's been there.

The department's initiatives to encourage participation beyond doctoral studies have included, as detailed above, the introduction of a maternity leave policy for post-doctoral researchers, careers talks from a diverse range of speakers and unconscious bias training. Despite these initiatives, staff acknowledge that a drop off at post-doctoral level continues. When we consider that female students outnumbered males in A-level science entries for the first time in 2019 (Weale and Larsson 2019), the high levels of female drop out suggest that the so-called 'leaky pipeline' persists, even in a best-case scenario department.

Widening Participation Policies

An important ASC emphasis for the department is to widen participation for disadvantaged students and extensive efforts have been made in that direction. In the context of rapidly increasing numbers in HE, and national policies that emphasise the inclusion of non-traditional student groups, Marsden, and the chemistry department specifically, have worked closely with schools to build bridges with potential students from socio-economically disadvantaged backgrounds. As Schools Outreach Officer, Audrey has aimed to recruit young people from diverse backgrounds in an attempt to make the chemistry department at Marsden a more inclusive subject area.

Audrey told us that once she understood the aims of the ASC it was clear the charter would fit well with the aims of the department. An important priority was the department's commitment to diversifying the staff body. She explained that

> Once I sort of knew what it was about, I realised that I had been working along its goals, and aims, and ambitions, for some time. Because, I'd been involved in other initiatives, national initiatives, trying to encourage women in science, and people from black and minority ethnic backgrounds into science.

Audrey worked to widen participation in science subjects before the department gained the ASC and this success enabled her to consolidate her efforts in that direction. Her previous experience showed how difficult it can be to change non-traditional students' perceptions of studying science and convince them the subject is interesting and attainable.

Audrey's work has increased since the ASC award and several new initiatives were introduced to improve the diversity of the undergraduate student population. The department has a partnership with a sixth form college in London with a majority population of minority ethnic students. Audrey explained that within the school the general feeling of staff is that their students do not go to university. Audrey frequently visits and runs workshops to try and engage the students by raising their aspirations and encouraging them to consider the possibility of attending university. She also selects a small number of the students to attend Marsden at the start of their school term so they can have a sense of what the institution feels like.

When choosing geographical areas suitable for widening participation activities, Audrey said she is 'choosy' about where she goes: 'I look at the postcode thing, as well as how well off people are, which you look at free school meals, and things like that. So, actually, a lot of the rural schools, there has been no tradition of going to university … perhaps it's not been the ethos, necessarily, to go to university, so going to some of the little rural schools is important.'

These efforts to engage non-traditional students were noted by other members of staff, together with the challenges of engaging them in HE. Susie commented that 'it seems that those ethnic groups like to stay very close to home. And even, you know, being half an hour away … can be a big barrier in some ways.' Like Audrey, Susie was aware of the issues associated with attempts to widen participation to include non-traditional student groups. Minority ethnic students, particularly females, are more likely to live at home and commute to campus, for a range of cultural reasons (Chappell et al. 2018). The difficulties of engaging non-traditional students in HE are reflected in DfE (2018) figures, which confirm the extent to which income inequality persists and social background continues to influence students' choices and pathways.

Social Class Representation in the Department

Despite significant work aimed at improving staff diversity, the social class background of staff and students in the department is, according to Hayley, 'overwhelmingly middle class.' Audrey feels that social class is less visible in the lab because 'students tend to dress fairly uniformly, particularly in the chemistry department, where you're fairly restricted on what you can wear.... So, it does homogenise people quite a bit in that way.' But Linda had a different perception:

> I'm not sure that class is invisible in the chemistry department. I would say that it's very noticeably white middle class here... Particularly, undergraduate recruitment, it wasn't what I expected when I came to university, put it that way. I thought it would be more diverse, and there would be more, you know, people with coloured hair and piercings, there was not any of that at all. So, I think it does have a feel of being a very middle class university.

Robin also viewed the department as being predominately middle class and Rachel concurred and suggested the department 'is very white, middle class.' Noah commented that 'Social class ... that's a hard one for us Britons to get around.' He said social class is difficult to address 'from the context of a change that we could make as a department. And I don't mean to say therefore we should do nothing. But I do think that ... they're intractable problems. That's not to say insoluble.' Work on the ASC has yet to make a significant difference to the representation of the student or staff body, except in relation to gender representation. It may be, however, that the department's policies may achieve a more representative student and staff body over time.

Linda recognises the obstacles to progress:

> The problem is that, the middle classes are always going to want to come to university, and so you're always competing against that, against that privilege, and that additional support. And, that's what's difficult is you're always going to be disadvantaged at the start, so you're always going to need that extra grit to work your way up and to compete against that.

Audrey's acknowledgement of the disadvantages faced by non-traditional students is well documented in existing research (Waller et al. 2015; Reay et al. 2009). As Reay et al. (2009: 1116) noted, a significant failure of 'the widening access and participation debate' is the lack of recognition that 'elite universities need non-traditional students just as much as the students need them ... Within the current status quo, an enormous number of working class students are excluded from realizing their academic potential.' Moreover, Reay et al. suggest that this contributes to elite institutions lacking 'a rich social diversity,' which diminishes the university experience for all. We found evidence of limited social diversity in the case study, despite Athena Swan efforts to foster inclusion and diversity. Archer and Tomei's (2013: 11) TISME work revealed that improving the representation of girls, minority ethnic and working-class students in science is a complex task.

> Liking science is not enough. Many children (especially girls) rule out STEM choices even though they like science, are good at it, and think what scientists do is valuable. Children, young people and their families have a narrow view of the careers that STEM qualifications can lead to. Many think in terms only of the obvious and traditional routes (scientist, engineer, doctor) and are unaware of the wide range of careers that STEM qualifications make possible.

To encourage greater participation from under-represented groups, Archer and Tomei (2013: 11) suggest that children 'need to be informed that science and mathematics qualifications can 'keep options open' and lead to a range of well-paid and interesting careers both within and beyond STEM.'

The patterns of recruitment and participation in STEM and STEM-related subjects within the department have been relatively stable in terms of students' social class backgrounds and ethnic groups. The department's effort to implement policies that have aimed to widen access are yet to make a significant impact on attracting young people from less socio-economically advantaged backgrounds to study at Marsden. Russell Group universities have concentrations of non-disadvantaged students that are designed into their selective role and status, and this defeats attempts to recruit more widely. Our case study confirms that the occupational class distribution of students in the department remains relatively unchanged and that chemistry is overwhelmingly studied by middle class students with family members who have a STEM background (Smith 2011).

Conclusions

Marsden's chemistry department has made significant progress in improving the working environment through the ASC awards process. However, we have also noted in this chapter that prior to discussions about the ASC taking place, the department was described by all staff as having an inclusive and enabling ethos and was therefore already aligned with the core aims of the Charter. Flexible working arrangements, shared parental leave and a maternity leave policy for PhD and post-doctoral researchers have all been introduced in response to the department gaining the Athena Swan award and have only enhanced the already positive working environment. These various initiatives and policies were recognised by participants as improving the work life balance of staff and students in the department. Those staff with caring commitments felt they particularly benefited from the introduction of the ASC.

Yet despite the worthy policy activity and the various associated initiatives discussed in this chapter, the staff feel the department still needs to make significant further changes to achieve greater inclusiveness in the make-up of the staff and student bodies. Some of this work needs to happen before the sixth form, particularly efforts to improve the limited social class and ethnic diversity in the department and in other Russell Group universities. The balance between selection and inclusion is unlikely to be resolved at departmental level. The department can only recruit students from the available pool and if this pool is unrepresentative, then recruitment will reflect the inevitable bias. This finding is consistent with Archer et al.'s (2018: 746–747) study that argues that existing conditions in schools make it 'very difficult for many students, but particularly girls and "quiet" boys from working-class and minority ethnic communities, to perform and assert an intelligible "voice" in classroom science.' Without more enabling conditions in schools, it is going to be hard for the department or any department to diversify its student cohort in line with government expectations. With few available non-traditional students with the right qualifications for the course, the department's efforts can only achieve a limited impact.

The academic nature of the chemistry course, described in Chap. 3 as intensive and demanding by some of the students, presents a further challenge to diversifying the cohort. Even amongst our participants, who are all academically capable and possess much of the necessary forms of cultural and science capital, we noted a few students who were struggling with the workload and who acknowledged that to be a committed chemist leaves them with little time for anything else. The dispositions and science capital required by students to succeed in chemistry at undergraduate level and beyond will present challenges to the department as it seeks to diversify the cohort by attracting more minority ethnic and working-class students (Archer et al. 2012). Where children do not possess reified forms of cultural capital and science capital, there are few opportunities to accrue this type of knowledge and acquire the associated class confidence through the English state education system (Reay 2017; Belmi et al. 2020).

The data leads us to argue that institutions should challenge the socially constructed identity of the scientist as white and male; and to recognise that female, minority ethnic and working-class students are equally capable of becoming reputable scientists (Archer and Tomei 2013; Burr 2015). Otherwise the dominant constructions of the capable, 'natural,' male scientist will continue to hamper policy efforts to increase inclusion. Without a strong and stable chemistry 'pipeline,' capable of supporting and enabling students at every level of their education, from primary school onwards, little is likely to change in HE and beyond.

One of the most encouraging areas of change in the department relates to improved female representation since 2007. An already positive attitude and atmosphere around gender issues has only improved since the department gained Athena Swan recognition. The main finding here, however, is that while gender equality has improved, particularly up to post-doctoral level, there is very limited diversity by social class or ethnicity. According to participants, widening chemistry participation to minorities and students from working class backgrounds remains a work in progress.

Proposition three is supported, therefore, by the evidence reported above. Gender and ethnicity do have a major influence on self-identity and aspirations to study and choose a career in science. Proposition four is also confirmed as broad issues of identity have a stronger influence than equal opportunities policy, and this explains the very slow progress towards a more socially inclusive chemistry department. Noah, however, remains optimistic about the potential for change, pointing out that whilst improving representation in terms of social class and ethnicity would be a challenge, 'we love hard problems, so you know, we should be able to come up with something.'

References

AdvanceHE. (2019). AdvanceHE's Athena SWAN Charter. Retrieved February 1, 2020, from https://www.ecu.ac.uk/equality-charters/athena-swan/.

Archer, L., & Tomei, A. (2013). *What Influences Participation in Science and Mathematics? A Briefing Paper from the Targeted Initiative on Science and Mathematics Education (TISME)*. King's College London.

Archer, L., DeWitt, J., Osborne, J., Dillon, J., Willis, B., & Wong, B. (2012). Science Aspirations, Capital, and Family Habitus: How Families Shape

Children's Engagement and Identification with Science. *American Educational Research Journal, 49*(5), 881–908.

Archer, L., Nomikou, E., Mau, A., King, H., Godec, S., DeWitt, J., & Dawson, E. (2018). Can the Subaltern 'Speak' Science? An Intersectional Analysis of Performances of 'Talking Science Through Muscular Intellect' by 'Subaltern' Students in UK Urban Secondary Science Classrooms. *Cultural Studies of Science Education*. https://doi.org/10.1007/s11422-018-9870-4.

Bagilhole, B. (2007). Challenging Women in the Male Academy: Think About Draining the Swamp. In P. Cotterill, S. Jackson, & G. Letherby (Eds.), *Challenges and Negotiations for Women in Higher Education* (pp. 21–32). Dordrecht, The Netherlands: Springer.

Belmi, P., Neale, M. A., Reiff, D., & Ulfe, R. (2020). The Social Advantage of Miscalibrated Individuals: The Relationship Between Social Class and Overconfidence and Its Implications for Class-Based Inequality. *Journal of Personality and Social Psychology, 118*(2), 254–282.

Buckingham, J. (2019). Athena Swan Review – Update from the Steering Group, AdvanceHE. Retrieved January 1, 2020, from https://www.advance-he.ac.uk/news-and-views/Athena%20SWAN%20Review%20-%20update%20from%20the%20Steering%20Group.

Burr, V. (2003). *Social Constructionism* (2nd ed.). East Sussex: Routledge.

Burr, V. (2015). *Social Constructionism* (3rd ed.). East Sussex: Routledge.

Chappell, A., McHugh, E., & Wainwright, E. (2018). *Successful Students: Exploring the Factors That Encourage and Enable Students from a Widening Participation Background to Stay the Course (Report)*. Uxbridge: Brunel University London.

Crozier, G., Reay, D., James, D., Jamieson, F., Beedell, P., Hollingworth, S., & Williams, K. (2008). White Middle-Class Parents, Identities, Educational Choice and the Urban Comprehensive School: Dilemmas, Ambivalence and Moral Ambiguity. *British Journal of Sociology of Education, 29*(2), 61–272.

Department for Education (DfE). (2018). *Widening Participation in Higher Education, England, 2016/17 Age Cohort – Official Statistics*. GOV.UK.

Greenfield, S., Peters, J., Lane, N., Rees, T., & Samuels, G. (2002). *Set Fair, A Report on Women in Science, Engineering, and Technology*. London: Department of Trade and Industry.

Hoskins, K. (2012). *Women and Success: Professors in the UK Academy*. Stoke-on-Trent: Trentham Books.

Le-May Sheffield, S. (2004). *Women and Science: Social Impact and Interaction*. New Brunswick, NJ: Rutgers University Press.

Reay, D. (2017). *Miseducation: Inequality, Education and the Working Classes*. Bristol: Policy Press.

Reay, D., Crozier, G., & Clayton, J. (2009). Strangers in Paradise: Working-Class Students in Elite Universities. *Sociology, 43*(6), 1103–1121.

Smith, E. (2011). Women into Science and Engineering? Gendered Participation in Higher Education STEM Subjects. *British Educational Research Journal, 37*(6), 993–1014.

Smith, E., & White, P. (2011). Who Is Studying Science? The Impact of Widening Participation Policies on the Social Composition of UK Undergraduate Science Programmes. *Journal of Education Policy, 26*(5), 677–699.

Smith, E., & White, P. (2018). *The Employment Trajectories of Science Technology Engineering and Mathematics Graduates, Final Report*. Leicester: University of Leicester/University of Warwick.

Waller, R., Harrison, N., & Last, K. (2015). Building a Culture of Participation: Interviews with the Former Directors of the National Aimhigher Programme. Retrieved June 3, 2019, from https://uwe-repository.worktribe.com/output/829528/building-a-culture-of-participation-interviews-with-the-former-directors-of-the-national-aimhigher-programme.

Weale, S., & Larsson, L. (2019, August 16). 'Students Don't See the Value': Why A-Level English Is in Decline. *The Guardian*. Retrieved December 3, 2020, from https://www.theguardian.com/education/2019/aug/16/students-dont-see-value-a-level-english-decline.

CHAPTER 5

The Limits of Equality Policy

Abstract In this chapter, we draw on the interviews with staff to examine areas where the ASC still needs to lead to changes. We have identified four key areas where additional forms of support could further enhance equal opportunities. First, improved support for part-time staff applying for promotion is needed. Second, departments should challenge the culture of overworking as the dominant model for achieving career progression. Third, greater representation of LGBTQI+, women, ethnic minorities and less advantaged social groups within the staff and student bodies is necessary. Fourth, is the need to widen participation to include non-traditional, working class and disadvantaged students.

Keywords LGBT • Ethnic minorities and gender

The department's journey with Athena Swan followed years of growing support for equal opportunities, both at departmental and university levels. During its work with Athena Swan, the department was assiduous in producing action plans to comply with charter criteria and to evidence and report progress. The first action plan was produced in 2007 and identified five key areas for development:

K. Hoskins, B. Barker, *STEM, Social Mobility and Equality*,
https://doi.org/10.1007/978-3-030-49216-8_5

1. Baseline and science, engineering and technology (SET) academic profile.
2. Positive support for women at career transition points.
3. Changing the culture and gender balance in decision-making.
4. Work life balance practices—introduction and uptake.
5. Champions, responsibilities and accountabilities.

These five priorities continued into the next action plan, published in 2010. Within each of these priority areas, however, there were changes to reflect the introduction of new policies and practices. In 2007 for example, within point 1, early attention was given to 'Increasing the percentage of female academic staff in different categories.' By 2010, that issue was refined to become: 'Continue to increase the percentage of female academic staff in all categories. Over a 6-year period we have increased from 8% to 32% of senior staff being female, with 22% of staff at professorial level being female as of May 2010.' This chapter draws on staff interviews to capture their reflections on areas within points 1–5 above where policy changes and improvements are needed.

The perceptions of staff participants detailed in Chap. 4 capture the success but also the disappointments and dilemmas that have arisen from their determined efforts to reduce the impact of class, gender and ethnicity on students' choices, academic progress and careers. The AS Charter has provided a framework and impetus for further developing the inclusive culture and ethos of the department. Staff and students (in Chap. 3) alike describe their sense of belonging to a considerate, caring and inclusive culture that has grown stronger over time. The benefits have been experienced by almost everyone, with issues around gender and sexual orientation no longer a barrier to full participation. There is great support for parental leave and there have been major initiatives, including unconscious bias training, to eliminate all forms of discrimination. Women are well-represented amongst undergraduates but less well so amongst post-doctoral students and lecturers.

Four areas have emerged from the case study, however, that are in need of further action to enhance equal opportunities. First, there is a need for improved support for part-time staff applying for promotion. Second, departments should challenge the culture of overworking as the dominant model for achieving career progression. Third, greater representation of LGBTQI+, women, ethnic minorities and less advantaged social groups within the staff and student bodies. Fourth, is the most difficult of all, the need to widen participation to include non-traditional, working class and disadvantaged students. The key dilemma is that Russell Group universities are by definition elite and selective. The Marsden chemistry department is 'overwhelmingly middle class.' It seems unlikely that this strong recruitment bias can be corrected at departmental or even single university level, however enlightened their administrators, professors and colleagues.

The issue is compounded by the limitations of the Equality Act 2010, which is proving ineffective in tackling the under-representation of those from varied social class backgrounds. The evidence here suggests that the weakness of the 2010 Act stems from the working of its protected nine characteristics. These characteristics do not recognise adequately the intersectionality of multiple identities and the synergy between the protected characteristics. Social class, family capital and inequality continue to be under-estimated as influences on identity, aspiration and the ability to access opportunity.

Promotion Processes

Staff regard the promotion process for part-time staff as an important area for further development. As indicated in Chap. 4, the department's flexible working initiative was cited as a flagship scheme and was mentioned in all the staff interviews. Interviews referred to some of the practical issues involved in implementing the scheme, but the conclusion was that the department is large enough to absorb most of the extra workload.

The flexible working initiative created dilemmas for staff on fractional posts who applied for promotion. Tensions stemmed from the number of publications and research grant applications required for the fractional post held. Linda was involved in recent work to ensure that the 'promotion criteria are fair, and that the factoring in of part time working and career breaks, will be taken into account.' Linda explained that 'in theory, the promotion criteria have just changed, this is the first year that we've been through this new process, and I don't think we'll know for a while whether it has made a difference.' However, she said the key point was to ensure that 'part time working should be accounted for when calculating, for example, your research income, or your number of papers.' But Linda acknowledged 'that's the theory, whether that pans out, I don't know!' The staff were aware of the need to ensure that any ASC policy change or initiative introduced did not inadvertently create inequality in relation to other aspects of work.

Rachel pointed out that part-time contractual hours are supposed to be taken into account 'when you do your appraisal and you go for promotion,' so that if you are part-time working, it needs to be factored into staff workload models, so you shouldn't be given too much to do so your research is affected.' Susie also discussed the tensions produced by the part-time working initiative, and recognised that the university needed to take a role in resolving tensions arising from the promotion criteria. She explained that

> I think even in terms of promotion, there is work needed from the university's point of view. They seem confused about what they expect from part-time workers. So you're very much allowed to go for promotion if you work part-time, but there doesn't seem to be a clear consensus what it should mean. So if you work .8, does it mean you have to achieve .8 of all the criteria, or does it mean you have to just achieve .8 of the criteria? And that actually makes a difference, and that's something that at the moment they haven't really got to grips with.

Susie raised some important equality issues in relation to this point. Flexible academic working has created significant tensions, especially for part-time staff (Morley and Walsh 1995; Manfredi and Holliday 2004; Toffoletti and Starr 2016). The lack of recognition amongst promotion panel members for staff on part-time contracts represents a significant issue at institutional level. The lack of clear guidelines about if and how to mitigate expectations for part-time staff is a key challenge, not only for promotion panels but more broadly in the STEM sector. The departmental view is that there is more work to do on this issue, despite gaining the ASC award, suggesting some weaknesses in the ability of the ASC to bring about policy changes to resolve this issue.

As Susie noted, a further concern about promotion is the different standards needed within grades. Susie and her colleagues were recently involved in looking again at promotion 'not just promotion to professor but promotion within the grades of professor, which is quite a big issue, isn't it?' There is a significant difference in the workload profiles of high-status and low-status professors. Leathwood and Read (2012: 17) consider these differences. One of their participants refers to professors on differing grades as 'racehorses and carthorses,' as a way of describing the distinctions that exist within the professorial grade. The racehorses spend the majority of their time on funded projects, the carthorses have to pick up a larger teaching and marking load to compensate.

Despite the difficulties, however, Robin feels there is hope of progress with this issue. One of the department's 'mega star research professors … is eighty per cent time, has been for years and … is very well respected as a researcher.' Robin explained that this particular professor had achieved the highest level within the professorial band and this sends a clear message for people to see 'that it can be done. And she's always been judged on the basis that she's working eighty percent of the time.' This positive example of a very capable woman making a significant contribution to the intellectual work of the department, whilst managing her care commitments, provides a strong role model for the women around her. There are dangers however, in using outstanding female scholars as professional benchmarks.

The pressures of the job, particularly institutional demands for grant capture and publications, is a key issue for part-time and full-time academic staff, in the department and more broadly. These pressures led several staff participants to overwork, an increasing problem for many academics.

Academic Culture of Overworking

As reported in Chap. 3, some members of the student sample were struggling to manage their undergraduate workload. Staff also commented that the dominant work culture, not only in the department, but also as an established norm within the sector, requires them to overwork to gain promotion.

Hayley noted 'there's still very much a long hours culture' here. The department does provide opportunities for staff to change their working patterns or to reduce their hours but Hayley feels 'there's still really underneath, this expectation that you will work more than the hours you get paid for.' As she remarked: '…if you've got that flexibility to go off and do what you need to, that work doesn't go away and it's still got to be done at some stage. And that when we reduce somebody's hours, we really don't probably reduce the role enough to compensate for that.' Hayley felt that in practice, the flexible working policy simply shifted the work around, and staff were generally expected to pick up each other's workloads, often at short notice.

Robin experienced similar pressure to overwork, particularly after he adopted his child and reduced his hours to 0.8. He explained that 'I know that having done part-time working, there are all sorts of issues.' His main concern was overwork: '...before I was part-time working, I was probably working one hundred and forty percent of workload. And once you're actually only being paid to do a certain number of hours a week, you feel, "Well, why should I really be doing any more than eighty percent of my workload?"' Questions about how much work to do were directly linked to going for promotion and worries about the fair distribution of work across teams.

As Director of Undergraduate Teaching, Robin recognised the challenge of distributing work equitably. He explained that managing academic workloads can be difficult with full-time staff and 'it can be very difficult to balance workloads when staff are part-time working.' Robin acknowledged that this is not just an issue in the department but also 'across the whole of academia.' As he noted, 'there's people working way beyond, as standard, the contracted hours. So, managing part-time working within that context so that you're being fair to the person who's part-time working and fair to other colleagues, is quite difficult actually.' The issue of overworking emerged from the interviews as a significant concern, with the lack of wider institutional support clearly an area where work with the ASC should go further to protect staff and ensure equity in the distribution of activities, for all staff, whether on part-time or full-time contracts.

Diversifying Chemistry

According to participants, the ASC has created a positive context for improvement but the good practices identified in the department have yet to reach across all the nine protected characteristics of the Equality Act (2010). Participants recognised the need for further diversification of the student and staff body to achieve full representation of LGBTQI+, women, ethnic minorities and those from less advantaged class backgrounds.

LGBTQI+

One of the issues involved in supporting LGBTQI+ members of staff and students is the invisibility of these individuals in a university science department. To better understand the specific difficulties facing LGBTQI+ students, Robin explained that a survey was carried out about four years previously to understand the undergraduate experience in the department for those who identified as LGBT. The aim of the survey was to 'determine whether there were any problems that students were facing. Whether there was any bias that they felt or any bullying either from staff or within the student cohort.' The staff were able to use the feedback, which was generally encouraging, to implement recommendations for change. One issue noted was the lack of 'specific support from the department for LGBT students. So, we now have a list of staff members who are LGBT and allies, which is on the website, who are happy to discuss LGBT specific issues with students that might have them and not want to talk to their supervisor.' This approach was an attempt to evidence the 'breadth of support and allies in the department and a number of people that were more than happy to talk about these issues.' This support was welcomed by staff and students.

This approach to providing supportive staff contacts was acknowledged by Hayley who commented: 'we've got a number of staff who have said that they're happy to talk to individuals about LGBT issues, and we've made that … you know, the information available, and said, 'These are the staff members who are very comfortable talking to you about anything to do with this.' Rachel reported that there is 'good LGBT support now as well, I think they're really trying to focus on that to make sure that there's equality there.' Staff acknowledged that these efforts were a good starting point for supporting their LGBT students, but that more work was required, particularly in terms of challenging the stereotype of the sort of person that becomes a chemist.

Robin talked at some length about LGBTQI+ staff and students being 'hidden' in the department. He explained that 'one of the seminars for our graduate students talked about sexuality and science as a broader issue, because it's very much a hidden diversity issue in science.' Robin said one reason for this may be that there are 'very few out scientists at national or international levels. So, it's not recognised widely' and because of this lack of wider recognition, science and chemistry in particular are 'not considered a stereotypical career for people who are LGBT, and there still are stereotypical careers sadly that people imagine you're going to be a hairdresser or work in design or fashion or music or creative arts.' In Robin's experience, working in science means that 'if you're a gay male and if you're a gay female there's a whole different set of stereotypes that people pin on you.' For these reasons, Robin feels that LGBTQI+ issues are not recognised and supported as much as they could be, a perspective echoed by some of our student participants. George and Ainsley both noted limited support networks in place. The support networks on their own would be insufficient to break down and challenge the lingering bias and prejudice these students often experience.

Our findings confirm the National Union of Students (NUS) (2014) report exploring LGBT students' experience in HE. The report found that 'LGBT societies appear to play a major role in LGBT applicants' choice of university and in providing much-needed support to these students once they are at university' (NUS 2014: 47). The value of these groups was their ability to provide a 'supportive and non-judgemental environment,' which is 'crucial in enabling students who want to come out to do so without fear of discrimination or bullying' (NUS 2014: 27). However, there were exceptions to this pattern, with some participants in the NUS study feeling support groups were inadequate in meeting their complex needs. Whilst they appreciated the networks they could access, if they had not already come out, many felt they would struggle to do so once at university because of the persistence and impact of stereotyping (NUS 2014).

Our findings confirm this perspective. Challenging stereotypes is arguably even more pressing in STEM subjects where minority groups continue to be under-represented. Staff and students alike argued that more could be done through the ASC to challenge stereotyping of LGBTQI+ individuals seeking a career as a chemist.

Gender

The department has made significant strides in providing a working environment and culture that prioritises support for women. The ASC initiatives and policies have made a real difference to the careers of women in the department, not least through increasing the representation of women. The department has more senior female staff than many other comparable chemistry departments (in terms of size and resources), and this work has been enabled by the tireless efforts of the staff we interviewed.

But there was a recurring theme in the interviews about the disproportionate amount of caring duties for children, elderly relatives or extended family members undertaken by female staff and students. Susie explained: 'what we don't know is the amount of time that men in relationships spend in childcare or in-house care or in elderly parent care, whatever these demands are that tend to be placed on the females in the workplace.'

The values and goals underpinning decisions about career choice are, for many women, shaped by their family and caring commitments (Cotterill et al. 2007). The balancing act between work and home life can place women under considerable pressure. David and Woodward (1998: 213) found that 15 of their 16 respondents had married and 1 of the 16 had children. This led them to conclude that gendered caring commitments were a key factor for their cohort: 'nearly all the stories illustrate the fact that women, whatever their circumstances, have a period of their lives involved exclusively with family matters, mainly but not invariably, with children.' These periods of family commitments 'often preceded developing an academic career' (David and Woodward 1998: 213). In addition, it seems that caring commitments are just one ingredient in the gender inequality that women academics routinely experience in trying to achieve career progression (Santos and Dang Van Phu 2019).

Susie was aware that the sources of gender inequality were wider than just caring commitments and related more broadly to gendered imbalances in career pathways. But she also noted these issues are bigger than the institution, commenting: 'these are things that need changes in society.' She feels the only way it can change is by women working on a personal level to achieve equity at home. But she also felt such changes do not 'happen overnight. So, I think there are longstanding issues to do with that lack of equity that will have an impact.' Our data confirms existing studies and suggests that even where the ASC is embedded and is having an impact for some female staff further changes are needed.

There is a clear tension between STEM policies that appear to offer individuals the scope to achieve intragenerational social mobility whilst maintaining a healthy work life balance; and the lived reality in HE of an academic culture that tends to reward overworking as a means to promotion (Mavin and Bryans 2002; Hoskins 2012). Robin felt that even in a high performing department with extensive institutional support, chemistry is not especially woman friendly. What needs to change, Robin feels, is the wider culture surrounding being a female academic working in chemistry:

> The other barrier that still remains on the gender side, you know, it's just that continuing work with PhD students and post-docs, for them to feel that the career is one they want, and for them to see role models that is not people standing up and saying, 'Well the way I made a success of academia was by working ninety hours a week and sacrificing every semblance of a

> social life and personal life.' Which is sadly still what lots of academics do at conferences, you know. I go and meet people at conferences and the general thing is boasting in the bar about how many hours they've been working and how many papers they've been writing, and how many grants they've done, and how many conferences they've been to. And I can totally understand why a young student, be it male or female, who wants a family and wants to fully engage with that, says, 'Oh, can I do that with this job and listen to all these old guys boasting about, "I spent ninety hours a week doing this"'.

Our participants felt the culture of overworking was something they were likely to encounter in the department and at conferences. We noticed a subtext, an unspoken yet taken for granted norm, that to gain internal or external promotion, significant amounts of additional work are required to publish high quality outputs, and these demands of the job were off putting for junior colleagues.

This finding is confirmed by Canetto and colleagues who reveal that whilst participants wanted to have a satisfying career and family life, the graduates in their study deemed an academic career path in a science subject to be 'too demanding in terms of time, as well as too rigid in terms of its advancement structure' (Canetto et al. 2017: 19). Challenging and changing deeply embedded institutional cultures and institutional habitus in relation to gendered inequality is a complex, resource intensive task. Such masculine cultures tend to operate at submerged levels and can easily be ignored by senior leaders, particularly in a university where the policy context appears overtly supportive. For those students from working class, female and/or minority ethnic backgrounds contemplating a career in academia, the prospect of engaging with a hyper-competitive individualised work culture could be the key feature that tips the balance away from the academy and into a career in industry. Whilst masculinised cultures are present and well documented across many disciplines, STEM subject areas may compound and exclude female students in greater numbers than do the social sciences.

Robin spoke of a further, discipline specific issue about working in science. He remarked that 'in the sciences, it's considered a real weakness to show any aspects of personal life in work, because you're supposed to be totally objective, totally scientific about everything.' Robin has experienced how negative this view point can be, pointing out that it does not matter 'if you've got some crisis on at home or, your wife has just had a baby, or your husband's doing this, that or the other. It doesn't matter. It's science ... so you get on with it.' Robin feels that this perspective is 'corrosive, and it's a science specific thing. And that's why Athena SWAN in science has been a particular big push, because it's made it hard for women particularly to contemplate lives outside.' These are pressures women scientists are likely to encounter in their working lives and help to account for the under-representation of women in science, despite all of the support in place.

Susie reflected on the department's ambition to achieve gender equity: 'I think we'd like to see equity at every level through the career structure... But, you know, we're not as high up, I think it's fair to say, in terms of the professorial appointments for females when compared to males.' Although the department has around 25 per cent female professors, which is above the national average of 2–6% in the natural sciences (Bagilhole 2007), there is still space for further gender equity in the department.

Reflecting on why gender equity might be hard to achieve, Audrey feels that it's because of the 'adage that women always think they're ... this is probably over generalising, but there's that imposter syndrome, and they think, "I'm not good enough for this" and because of a lack of confidence, they tend not to apply for promotion as quickly as their male colleagues.' Audrey feels that women are concerned that 'if they put their head above the parapet somebody will discover that they really shouldn't be here, therefore, they do not go for promotion,' in effect excluding themselves from the running. This issue is well documented by Sandberg (2013) who argues that women need to 'lean in' to advance their own careers, referring to a self-promoting technique that is similar to that used by men.

Ethnicity

The participants all talked about the homogeneity of the staff body at Marsden and interviews touched on how this issue could be tackled, from more inclusive student and staff recruitment, to support for promotion and progression. Susie feels that it has been 'more of a challenge for us, particularly in terms of recruitment' to attract BAME students. When asked why she felt this was an issue, Susie explained that there are 'some quite diverse populations that aren't too far away,' but relatively few candidates apply from these areas.

Robin, feels that 'as a university, and it's partly geographical and partly the nature of the university, but this university is weak on ethnic minority background.' Robin explained that ethnic minority students in neighbouring cities tend to choose the local, post-1992 institution: 'it's a challenge here because... it's not stereotyping to say ... the ethnic minority students ... stay at home in the local community.' We asked Robin why he thought this pattern persisted and he said that: 'in part, it's probably because they feel very comfortable there and they feel that they might be uncomfortable here.' This reluctance to travel far from home seems to account for the lack of minority ethnic representation at pre-1992 universities (Chappell et al. 2018). But a further reason could relate to the tensions between minority ethnic students' habitus and the institutional habitus of pre-1992 universities (Thomas 2001). To counter this view, Robin explained that staff try hard to ensure they 'send the right messages that they would feel comfortable here. We engage through the WP scheme in part, specifically with schools in some of the areas where there's high ethnic mix.' By continuing to work on widening participation initiatives the department hopes to extend the ethnic diversity of the staff and student populations.

Robin said that Chinese students disrupt the trend and tend to be well represented in the department. He explained that 'as a university we have a lot of Chinese students ... that's our ethnic base. So actually, probably raw numbers look quite good.' Robin also noted that they have some ethnic representation in the staff base: 'we have academics who are Japanese, we have an academic who's a Sikh UK background in terms of ethnic minority.' Robin also noted that the department has colleagues who are 'German or Polish or Bulgarian, or so on,' but recognised 'you wouldn't necessarily call them ethnic minority, but we have quite a geographically mixed staff.' Rachel similarly noted that 'we have quite a lot of nationalities but we don't have very many ethnic minorities really working in the department.' Hayley remarked that the department does not yet have 'a particularly strong ethnic mix.' To offer the study of chemistry to more ethnic minority students, Hayley feels the department should 'ensure that we're as welcoming and as open as possible, and that we offer those opportunities to as many people as possible, and then hopefully that will then sort of snowball effect.' Hayley commented that staff are very aware of the need to have diverse teams to improve the quality of 'the outputs that you gain, and that diversity of thinking and thought, and background really helps.' This staff recognition suggests that the department should aim to increase recruitment from candidates with minority ethnic backgrounds to meet Charter aspirations.

Robin feels a lot more work is needed to translate ethnic diversity into undergraduate recruitment from minority groups. Audrey commented that a key area for further change was outside of the university itself. She recalled a recent widening participation visit to a local school where she was trying to 'encourage black and minority ethnic women into science.' During the course of her discussion with the students, they explained that they would not consider applying to Marsden because 'you can't get your hair done, because all the fancy braids, and all these things, there wasn't anywhere [locally] where you could go and get your hair done, because it's a really specialist thing.' These concerns were very real to the students, who said they would only consider applying to institutions that have 'somewhere where you can get your hair done, where there's a much higher black population in the city.' Audrey commented that whilst the university could do so much to support students, it's not only about the institution but the surrounding area and proximity to culturally relevant services and goods.

However, Audrey also noted that this issue had previously existed in relation to Chinese students, who had 'initially said, "you can't go shopping in [name of town], because you can't buy the right ingredients, you'd have to go to [name of town]."' But that this had all changed because of a recently opened Chinese supermarket. Now students can 'buy the ingredients. It's things like, being able to buy the things that you want, and being able to get the services that you need, like having your hair done,' that in Audrey's experience make all the difference to the university's ability to attract minorities.

Lasting change cannot be achieved without attending to the intersectional elements of women's and men's lives. Without understanding how gender, ethnicity and social class intersect to produce particular aspirations and opportunities to study a STEM subject, little is likely to change. Rather, efforts to change the ethnic, classed and gender composition of chemistry is likely to remain piecemeal and disconnected as the policies in this area need more coherence.

The lack of minority ethnic representation in the staff and student body is a major concern and suggests that ASC aspirations have yet to be realised. This relative absence seems to confirm that propositions 3 and 4 are justified in asserting that social class, gender and ethnicity have a powerful influence on self-identity and aspirations; and work against policies and charters designed to promote genuinely equal opportunities.

Social Class Background

Amongst the most difficult issues to resolve is the intractable nature of social class background and its role in shaping students' dispositions and science capital (or the lack of it). At Marsden, despite much worthy outreach work and chemistry department policy initiatives, the staff and students are overwhelmingly from middle class backgrounds.

Audrey commented that understanding the difference that Athena Swan recognition had made to improving the diversity of students' class background was difficult. She explained that 'I don't know whether Athena Swan has made a difference, I don't know. And, actually, when they get here, it's not always obvious what their background is, unless they divulge things to you.' Audrey feels that the lab acts as a social equaliser because it can be harder to spot differences that relate to social class background. Audrey spoke of the homogenising context of the lab. However, reflecting on her work in schools, she did acknowledge that certain groups of pupils were less likely to display the dispositions to study chemistry. She told us that:

> Perhaps I notice more when I'm going out into schools, because school will be in a particular geographical area, which probably has drawn people from a particular housing estate, or whatever. And, you do get comments about, well, 'the university is not for me', or, 'I don't know what my dad would say if I decided to go, he wants me to go out and earn money.'

Audrey's experiences of the difficulties associated with trying to persuade young people from working class family backgrounds to consider entering HE, and study a subject like chemistry, are well known and the subject of much discussion (Smith 2011). The dominant construction of a scientist is some distance from the lived reality and embodied self that many young people from working class families experience.

Ellen also noted that in her experience, 'where there is that problem, where there isn't a way out of it, is because people have very low aspirations, educationally.' Ellen feels that because of these low aspirations, working class young people 'don't have a thing of ever leaving that area [where they live] to go to university, going elsewhere. And it goes round in circles.' Rachel also felt that recognising young peoples' family background and their associated aspirations helped support them. Drawing on her own experience of growing up in a working class family, she explained that

> I came from two working class parents, my mum was a librarian, my dad was an electrician, and my brother went to university, he's four and a half years older than me, and I was always clever at school so I always felt that it was something I could go to do, so I did. And they always pushed me ... but I had friends in school whose parents were a lot more, 'Oh, I don't think it's for you'. So some of them didn't go and I know some who are actually doing their degrees now which is much harder with a family. So I think a lot of it is to do with your family life and I think it's very true what they say about it's not for you and you could go home and say to your parents, 'I'd really like to do this', and then they say 'I'm not sure that's right for you', because they don't understand it.

Archer et al.'s (2012, 2015) work on the *Aspires* project confirms that the possession of cultural resources shapes family habitus and influences dispositions that relay and reproduce social class and inequality. Aspires 2 follows the sample through to age 19 and concludes that 'aspirations (are) profoundly shaped by social identities and inequalities, particularly in relation to intersections of social class, "race"/ethnicity and gender' (Archer et al. 2020: 6). Once acquired, dispositions, like habitus itself, are durable but not eternal (Ball 2003), but they are arguably not easily changed and have a strong influence on behaviour. If dispositions to study science are not inculcated at an early age in the family and are not then reinforced by the institutional habitus of the school, it is difficult for young people to develop aspirations to study a STEM subject in HE.

In reflecting on the challenges of opening up chemistry to young people from working class family backgrounds who may lack the aspirations and dispositions to study a STEM subject, Hayley commented that: 'many of us have grown up in traditional, middle class society ... attending universities, red-brick traditional universities where it's sometimes hard just to break those barriers down.' Hayley acknowledged the intractable nature of social class.

Audrey commented that financial concerns were another issue facing young people from a working class background and tended to deter them from considering HE. She explained that

> Actually, a lot of students are worrying about the debt issue, with the student loans, and the fees, and so on, which you end up with as a debt. But I

> usually point out that, actually, nobody ever pays it all back ... you probably won't ever be earning enough to pay it back within the payment window.

The policy storm arising from the introduction of tuition fees and subsequent increases in those fees in HE in England is well documented (Hoskins 2017; Wilkins et al. 2013). There is a marked aversion amongst many working class people to taking on debt for HE. One comparative study found that 'debt-averse attitudes remain much stronger among lower-class students than among upper-class students' and reported this was even more pronounced in 2015 than it was in 2002. They believed 'debt-averse attitudes seem more likely to deter planning for HE among lower-class students in 2015 than in 2002' (Callender and Mason 2017: 20). This is consistent with Audrey's perception that young people from disadvantaged backgrounds are not easily persuaded to overcome their inherited dispositions; and that the task is even harder when potential students are burdened with the extra worries generated by tuition fees and loans.

Linda said money is a real issue for young people from working class families in terms of 'people's infrastructure, in terms of the level of support that they get at home. The financial support, I mean.' On a pragmatic level, Linda feels students from middle class families have financial help from their families with weekly living expenses once they get on to a degree course. Working class young people have less or even no access to the Bank of Mum and Dad.

Robin reflected on trying to understand why the student and staff body lacks diversity. As leader of the undergraduate course, he was very concerned about improving inclusion. It was the one area that 'our admissions tutor didn't regularly report on. He reported on gender and A Level grades, and whatever. So, I got him to report with all our stats on, at the very least, the state school/private school breakdown of our intake.' They took the decision as a department not to 'delve deeper in to post coding and things like that. It's done centrally by the university, but as a department, we didn't have the resource to do that.' Their analysis of admissions showed the department 'were about eight three percent state school and seventeen per cent independent. Which is, the university as a whole is about eighty/twenty state school/independent.' Robin acknowledged that at first glance the 'figures look really good on paper, and it looks good for the department, for the university benchmark I think.' However, Robin pointed out whilst the figures look 'good as a base metric,' he is very aware from his widening participation work that as soon as post codes are included it does not look so good and that 'it's bad if you look at free school meals.'

He commented that the university does get 'a lot of state school kids... But they tend to be middle class, fairly upwardly mobile state school kids that have gone to good state schools and then they come here.' A more balanced picture of student origins could be achieved by using home post codes, eligibility for free school meals and the pupil premium as measures of undergraduate backgrounds. SMC reports indicate the difficulty of breaking through the class ceiling to open up HE to minorities and working class students (Friedman and Laurison 2020). Along with ethnicity, social class can impact on the likely pathways that young people will take through the education system in England.

The Limits of Equality? Rethinking the Equality Act, 2010

Despite gaining the ASC and implementing a range of policies and initiatives to improve equalities in this best-case example, we found that inequality persists at Marsden, particularly in the under-representation of staff and students from minority ethnic and/or socio-economically disadvantaged backgrounds.

The Marsden data shows that recruiting a genuinely inclusive intake continues to be extremely difficult, especially for Russell Group universities. There is a strong case for amending the Equality Act to explicitly protect those from less advantaged working class family backgrounds. Social class background, as it occurs in England, goes beyond the economic. It encompasses dispositions, norms and values that are steeped in class culture. These class cultures have evolved over time and in different geographic locations, leading to established class communities with their own distinct characteristics. The rich tapestry of social class in England makes it difficult to speak of simplistic social class groupings neatly bounded by income alone. Until the socio-economic duty provided for in section one of the 2010 Act is implemented and enacted, there is little chance of making any further progress with this issue.

But in the case of HE there is a much more powerful consideration. The way in which the sector has developed means that selective universities inevitability recruit students from a higher socio-economic background. Our discussion of the data in this chapter leads to the conclusion that social class is still the major influence on outcomes for everybody, and this is especially negative for minorities and the working classes. Therefore, the failure to bring in a provision of the 2010 Equality Act, socio-economic duty, suggests that policy-makers are not currently serious about equality. Until we have begun to implement that duty we have no chance of achieving further progress.

The evidence assembled here confirms proposition 5, as set out in Chap. 2. There is an intrinsic weakness in the Equality Act 2010 that stems from the operation of its protected characteristics. These do not and cannot recognise adequately the intersectionality of multiple aspects of identity and the synergy between the characteristics. Social class, family capital and inequality continue to be under-estimated as influences on identity, aspiration and the ability to access opportunity.

References

Archer, L., DeWitt, J., Osborne, J., Dillon, J., Willis, B., & Wong, B. (2012). Science Aspirations, Capital, and Family Habitus: How Families Shape Children's Engagement and Identification with Science. *American Educational Research Journal, 49*(5), 881–908.

Archer, L., Dawson, E., DeWitt, J., Seakins, A., & Wong, B. (2015). "Science Capital": A Conceptual, Methodological, and Empirical Argument for Extending Bourdieusian Notions of Capital Beyond the Arts. *Journal of Research in Science Teaching, 52*(7), 922–948.

Archer, L., Moote, J., MacLeod, E., Francis, B., & DeWitt, J. (2020). *ASPIRES 2: Young People's Science and Career Aspirations, Age 10–19.* London: UCL Institute of Education.

Bagilhole, B. (2007). Challenging Women in the Male Academy: Think About Draining the Swamp. In P. Cotterill, S. Jackson, & G. Letherby (Eds.), *Challenges and Negotiations for Women in Higher Education* (pp. 21–32). Dordrecht, The Netherlands: Springer.

Ball, S. (2003). *Class Strategies and the Education Market: The Middle Class and Social Advantage.* London: Routledge Falmer.

Callender, C., & Mason, G. (2017). Does Student Loan Debt Deter Higher Education Participation? New Evidence from England. *Annals of American Political and Social Science, 671*(1), 20–48.

Canetto, S., Trott, C., Winterrowd, E., Haruyama, D., & Johnson, A. (2017). Challenges to the Choice Discourse: Women's Views of Their Family and Academic-Science Career Options and Constraints. *Journal of Feminist Family Therapy, 29*(1–2), 4–27.

Chappell, A., McHugh, E., & Wainwright, E. (2018). *Successful Students: Exploring the Factors That Encourage and Enable Students from a Widening Participation Background to Stay the Course (Report)*. Uxbridge: Brunel University London.

Cotterill, P., Jackson, S., & Letherby, G. (Eds.). (2007). *Challenges and Negotiations for Women in Higher Education*. The Netherlands: Springer.

David, M., & Woodward, D. (1998). *Negotiating the Glass Ceiling: Careers of Senior Women in the Academic World*. London: Routledge.

Friedman, S., & Laurison, D. (2020). *The Class Ceiling: Why It Pays to Be Privileged*. Bristol: Policy Press.

Hoskins, K. (2012). *Women and Success: Professors in the UK Academy*. Stoke-on-Trent: Trentham Books.

Hoskins, K. (2017). *Youth Identities, Education and Employment Exploring Post-16 and Post-18 Opportunities, Access and Policy*. Basingstoke: Springer.

Leathwood, C., & Read, B. (2012). Final Report: Assessing the Impact of Developments in Research Policy for Research on Higher Education: An Exploratory Study. Society for Research in Higher Education. Retrieved December 31, 2019, from https://srhe.ac.uk/downloads/Leathwood_Read_Final_Report_16_July_2012.pdf.

Manfredi, S., & Holliday, M. (2004). *Work-Life Balance: An Audit of Staff Experience at Oxford Brookes University*. Oxford: Centre for Diversity Policy Research.

Mavin, S., & Bryans, P. (2002). Academic Women in the UK: Mainstreaming Our Experiences and Networking for Action. *Gender and Education, 14*(3), 235–250.

Morley, L., & Walsh, V. (1995). *Feminist Academics*. London: Taylor and Francis.

National Union of Students (NUS). (2014). *Education Beyond the Straight and Narrow: LGBT Students' Experience in Higher Education*. London: NUS.

Sandberg, S. (2013). *Lean In: Women, Work, and the Will to Lead*. New York: Alfred A. Knopf.

Santos, G., & Dang Van Phu, S. (2019). Gender and Academic Rank in the UK. *Sustainability, 11*(11), 1–46.

Smith, E. (2011). Women into Science and Engineering? Gendered Participation in Higher Education STEM Subjects. *British Educational Research Journal, 37*(6), 993–1014.

Thomas, E. (2001). *Widening Participation in Post-compulsory Education*. London: Continuum.

Toffoletti, K., & Starr, K. (2016). Women Academics and Work–Life Balance: Gendered Discourses of Work and Care. *Gender, Work & Organisation*. https://doi.org/10.1111/gwao.12133.

Wilkins, S., Shams, F., & Huisman, J. (2013). The Decision-Making and Changing Behavioural Dynamics of Potential Higher Education Students: The Impacts of Increasing Tuition Fees in England. *Education Studies, 39*(2), 125–141.

CHAPTER 6

Conclusion

Abstract In Chap. 6 we conclude by reviewing the book's two main purposes. The first is to use genealogical information about family circumstances, education and careers to better understand the formation of young people's identities and aspirations. The second is to review the implementation of chemistry department initiatives at Marsden University that have flowed from work with the ASC, and to consider their effectiveness in the wider context provided by the Equality Act (2010). In reviewing these purposes, we consider our case study data and published research to examine if these policies are convincing or effective. Where they are failing, we consider what can be done to improve the chances of success.

Keywords Socio-economic duty • Equality Act (2010) • Aspirations

This book has two main purposes. The first is to use genealogical information about family circumstances, education and careers to better understand the formation of young people's identities and aspirations. Our second is to review the implementation of chemistry department initiatives at Marsden University that have flowed from work with the Athena Swan Charter, and to consider their effectiveness in the wider context provided by the Equality Act (2010).

K. Hoskins, B. Barker, *STEM, Social Mobility and Equality*,
https://doi.org/10.1007/978-3-030-49216-8_6

We use qualitative data about three successive generations to make sense of the influences that inform our undergraduate chemists as they make critical decisions about future pathways. This leads us to question official mobility policies that have prevailed since the 1960s. These aim to expand participation and enhance quality throughout secondary and higher education, so that underprivileged groups can improve their lives and career progression. The intention is to increase the proportion of the workforce entering highly skilled occupations, especially in STEM-related disciplines, and to reduce wider inequalities. In light of this case study, and evidence from other publications, are these policies convincing or effective? If they are failing, what can be done to improve the chances of success?

Marsden & Athena Swan

The department's drive to improve equal opportunities for everyone, regardless of background, illustrates the tensions implicit in policy initiatives that seek to overcome disadvantage and promote fairness. Within the framework of the Athena Swan Charter, an inclusive culture has been established to enable everyone to enjoy high quality teaching and excellent opportunities for career progression. Participants are particularly appreciative of the department's values and culture. Marsden supports students and staff as they progress through the stages of careers in chemistry and this continuing work has earned recognition through prestigious national awards.

Equality and diversity are watchwords for the department, while a curriculum design that makes skilful use of industrial placements has opened many future possibilities. A sustained effort is made to tackle barriers to full equality. Staff participants are ready to 'acknowledge failure' and to identify the actions necessary for change, a procedure recommended by a recent critical review of Athena Swan and equality accreditation (Pearce 2017: 9). This is a 'best case' scenario for the policy proposition that HE institutions, reformed to match charter criteria, can pave the way for students to achieve their full potential, regardless of personal characteristics or background circumstances.

Marsden participants, especially the staff, are acutely aware of the barriers that impact on efforts to promote equality, inclusion and diversity. The chemistry department, like the university itself, is tangibly middle class, with the proportion of disadvantaged and ethnic minority students failing to match their representation in the wider HE population. Women chemists are represented as well as men through their undergraduate years but staff participants acknowledge the 'leaky pipeline' effect begins to operate from the post-doctoral stage onwards, with the pressures identified by Greenfield et al. (2002) apparently undiminished.

Chemistry at Marsden is fully subscribed but this is little consolation for the 21 per cent drop in the number of applicants nationally between 2015 and 2018 (McKie 2019). The obstacles to progress (for disadvantaged students, for women after the early stages of their careers) remain, despite sincere efforts to promote equality, and seem embedded in structures beyond the reach of even the most inclusive chemistry department, especially when very good intentions are compromised by a selective admissions strategy that aims to recruit only the most able students.

Five Propositions

This echoes the nationwide problem identified in Chap. 1. There has been an almost continuous swing away from science despite sustained efforts to promote STEM (Mandler 2017). Initiatives to help hardworking students access high quality opportunities, especially in STEM subjects, have yielded disappointing results. The government's own publications chart the impact of disadvantage on the chances of success at school, university and beyond (Social Mobility and Child Poverty Commission [SMCPC] 2014; SMC 2016, 2019). Educational policy has done much less to increase mobility than most politicians have assumed (Goldthorpe 2014: 431). Unequal mobility chances are persistent and can be extreme, especially in the case of long-range movement between routine wage-earning jobs and higher-level salaried positions (Bukodi and Goldthorpe 2018). Even well-qualified young people from less prosperous backgrounds, who secure professional employment with prestigious firms, fail to match the progression and income of their more advantaged peers (Friedman and Laurison 2020).

Inequality, Social Class and Family Habitus/Dispositions

Bourdieu's (1986) concepts, their use to interpret data from a national survey of children's science and career aspirations (TISME 2013; Archer et al. 2012) and our own previous work (Hoskins and Barker 2019), have guided our drafting of five propositions (see Chap. 2: 45) to explain why government policies have failed. The propositions led our analysis of the Marsden interviews and were confirmed by the participants' education and career trajectories (see Chap. 3). The mobility patterns at Marsden are rooted in and conditioned by family networks that transmit various forms of capital. Robin recognises Russell Group reality. The university gets 'a lot of state school kids' but they tend to be 'middle class, fairly upwardly mobile.' These students have been to 'good state schools' and 'then they come here.' There are multiple instances of habitus and dispositions operating through family networks to shape individual choices. As Thompson (1997) found with his hundred families, however, influences can be explicit, implicit or tacit and young people often choose destinations consistent with their own outlook and interests, rather than simply clone family history and habitus.

Family role models and resources are important nevertheless. Participants share similar backgrounds in financially secure, upwardly mobile, middle class families with abundant science capital. Nine of the 11 fathers and 7 of the 11 mothers have risen in the world compared with their own parents. Families act as a source of security for young people but also intervene to ensure the best outcomes they can, especially in education (Ball et al. 2000). This pattern resembles that found in the TISME (2013) sample. The Marsden chemists are able as well as eager to succeed in science, but their intellectual ability seems inextricable from their inherited dispositions and the capitals that facilitate high levels of support through education into the early years of a professional career. Their expertise in the STEM area, acquired through homes, schools and university, represents a further step forward for their families, with science capital renewed and deepened for the next generation.

Once set in the macro-picture described by Goldthorpe (2014) and Bukodi and Goldthorpe (2018), the Marsden students are less easily viewed through the government's mobility policy lens as a cluster of hard-working individuals prompted to STEM success by excellent schools. Instead, their interviews show family influences on expectations, choices and pathways; and illustrate how habitus and dispositions are transmitted through family networks as foundations for educational success. Our propositions follow Archer (2012) in asserting that inequalities in the distribution of capital, combined with differences in family habitus, produce class and race-based disadvantage. Social class has a major influence on self-identity, aspirations and outcomes (Reay 2017) while class origins remain 'sticky' throughout university and subsequent careers (Friedman and Laurison 2020). Education is not, after all, the great equaliser that breaks the link between social inequality and unequal opportunities (Bukodi and Goldthorpe 2018). Despite worthy efforts to promote equal opportunities, including the Athena Swan Charter and the Equality Act (2010), inequality, social class and families continue to shape young people's dispositions and choices. These influences operate across the social spectrum, prompting and also inhibiting young people as they absorb the world around them and consider 'what is possible or unlikely for a particular group in a stratified world' (Swartz 1997: 103) before tending towards social reproduction (Bourdieu and Passeron 1977).

Although Bourdieu's theories of cultural capital and social reproduction have been subject to extensive critical review, and certainly open as many questions as they resolve (Goldthorpe 2007), we have followed Archer et al. (2012) in deploying the concepts of habitus and dispositions to interpret the Marsden data. Goldthorpe regards the theory of social reproduction as 'quite unsound' but acknowledges nevertheless that in its 'domesticated understanding' Bourdieu's work is 'tolerably sound' and worthy of 'serious consideration' for research into 'the sources of social inequalities in educational attainment' (Goldthorpe 2007: 18).

The emphasis our participants place on the role of their antecedents and parents in guiding their attitudes and preferences, and in preparing them for education, especially in technical and scientific disciplines, suggests the importance of families in pre-conditioning responses to formal education. Bourdieu's concepts provide a framework for understanding why apparently equal opportunities in education lead to differentiated outcomes, and why repeated reforms fail to close the performance gap between disadvantaged students and their peers. The impact of social and family circumstances on young people's choices is greater than that of government policies designed to increase the proportion of upwardly mobile young adults seeking employment in the professions or in highly skilled STEM-style environments. Our case study data (see Chaps. 3 and 4) confirms that propositions 1–3 are valid. The empirical data does not support the popular belief that education is a great equaliser (Bernardi and Ballarino 2016).

Gender, Sexual Orientation and Ethnicity

Throughout the book, we have argued that social constructions of gender, gendered roles and gendered stereotypes have a notable influence on young women's self-identity and perceptions of what is possible for them in terms of subject options and vocational choice. Their careers are more likely than men's to be disrupted by marriage, family and caring responsibilities in ways that undermine fairness and equal opportunities (Chilisa 2006: 249). The characteristics protected by the Equality Act (2010) may also impact on young people's self-identity and often cause individuals to be treated unfairly, and as a result to be discouraged or limited in the pursuit of their aspirations and potential opportunities. As we have seen, the Marsden chemistry department has worked on gender, sexual orientation and ethnicity since 2007, developing initiatives within the ASC framework to ensure equity and inclusion.

Staff participants discussed parental leave (extended to doctoral students), flexible working around childcare to preserve careers, unconscious bias interview training, work on staff appointment practices to ensure less well represented groups are not overlooked, and diversity amongst visiting speakers and doctoral examiners. They acknowledge that progress has been less than they hoped. Only 25 per cent of the department's professors are female, the number of lecturers does not reflect the undergraduate population and there is a continuing loss of women from the university, mainly during the transition from the doctoral to post-doctoral stages. Flexible working is seen as helpful and inclusive but budget and staffing issues mean that it often ends up with colleagues picking up one another's workloads. Staff participants emphasise the disproportionate amount of caring duties for children, elderly relatives or extended family members undertaken by female colleagues and students. These invisible tasks have notable consequences for chemistry careers shaped by a long-hours culture and excessive publication requirements. Workload demands are seldom adjusted sufficiently to reflect part-time contracts.

Although gender representation has improved, there is little difference in the proportion of BAME undergraduates in the student body. The outreach officer works to diversify recruitment and has forged partnerships with schools to encourage BAME students to consider Marsden. Rachel drew a distinction between 'quite a lot of nationalities' working in the department with the fact that 'we don't have very many ethnic minorities' in chemistry. Although Susie said there are 'some quite diverse populations that aren't too far away' there are few applications from these multicultural areas. Robin suggested that 'it's probably because they feel very comfortable there and they feel that they might be uncomfortable here.' This acceptance of a low application rate seems to justify the Labour MP

David Lammy's comment that: 'Many universities spend significant bags of money on access, but can demonstrate little or no positive outcomes. Real progress in this area will require radical and punitive action by the government and Office for Students. There is clear market failure and this social apartheid is not in the interest of wider society' (Elgot 2018: unpaged).

Robin said that a Marsden survey concerning LGBTQI+ revealed 'it's very much a hidden diversity issue,' with male-based science and chemistry stereotypes 'pinned' on gay people. Support networks are in place and students welcome the inclusive culture of the department but they also doubt whether enough is done to challenge the lingering bias and prejudice they experience. Subtle cultural assumptions can increase the discomfort felt by those with characteristics that do not allow them to fit the dominant mould easily. They may limit what seems possible and discourage affected people from forming long-term career ambitions. Robin remarked: 'in the sciences it's considered a real weakness to show any aspects of personal life in work, because you're supposed to be totally objective, totally scientific about everything.'

Staff perceptions of the degree of progress with key issues, like gender representation (from the doctoral level onwards) and inclusion (of BAME and socially disadvantaged students), reported and discussed in Chaps. 4 and 5, confirm that gender, sexual orientation and minority ethnic backgrounds have an important impact on self-identity as well as on decisions to study and choose a career in science. Self-identity does have a stronger effect than equal opportunities policies that are supposed to counter the structures and stereotypes hindering full equality. Proposition 4 is therefore valid. Marsden chemistry is a leader in terms of Athena Swan, but progress towards a more socially inclusive department is slow. Underlying structures and attitudes remain a formidable barrier, even in a 'best case' scenario. Les Ebdon, former director of the Office for Fair Access, has a similar view and believes that although universities are making progress, especially on widening access, their proportion of disadvantaged students is far below target (Hall 2017).

Staff participants fully recognise the limitations of their work to ensure equity and inclusion. Many of their thoughts are consistent with points raised at a workshop on issues arising from Athena Swan, and other equality accreditation schemes, hosted by the Centre for the Study of Women and Gender at Warwick University (Pearce 2017). Delegates were clear that equality work should be fully recognised as work, so that an unfair burden does not fall on individuals, particularly women and non-binary people, and should be explicitly integrated into workload allocation. They argued that integrated support for intersectional equality work is necessary, and that women, men and non-binary people could be proportionally represented on Athena Swan self-assessment teams and equality committees. There was a strong desire for institutions and departments to build collaborative links, rather than directly compete, with resources, ideas and information on successful equality and diversity activities shared across HE (Pearce 2017).

Sociology's Inconvenient Truth

Our findings are consistent with other studies (Brown 2013; Friedman and Laurison 2020) that challenge the government's understanding of the structures, processes and people involved in social mobility. We believe it is wrong to discount sociology's 'inconvenient truth' (Brown 2013) in favour of an individualist model that exaggerates human agency and under-estimates the impact of relative poverty on young people's lives. Appendix 1 (Family Occupations and Class Identities) provides a complex micro-view of the ways in which successive generations within a family network build and transmit dispositions and multiple forms of capital. As in the TISME (2013) survey, these able, energetic, science-inclined young people are the result of an accumulation of relatively small advantages within their families, not products of a dose of cultural capital injected by teachers (Gladwell 2008). These same processes are at work for everyone, of course, with relative advantage and disadvantage transmitted in all their complexity in a great variety of families and circumstances.

Our five propositions help explain why it is so hard to overcome social class and family background and to conjure equal opportunities from a social fabric that is permeated with inequality and injustice (Wilkinson and Pickett 2009). The circumstances for disadvantaged families in the UK have deteriorated sharply in recent times (Health and Social Care Committee [HSCC] 2019). There is less support for every aspect of their lives than before, with New Labour initiatives, like Sure Start and Education Maintenance Allowances, either cancelled or allowed to disappear in successive waves of local authority spending cuts (O'Hara 2014; Barker 2010).

Staff participants involved in outreach activities do not seriously expect to find a ready supply of grade A scientists in POLAR 4, quintile 1[1] geographical areas. Robin said Marsden admissions look acceptable at first glance, but that 'as soon as post codes are included, it does not look so good.' Despite great efforts, including a programme organised by an experienced outreach officer, the admission profiles of the chemistry department and Marsden itself are similar to those of other Russell Group members. Students from the most disadvantaged fifth of UK POLAR areas account for just 6.2 per cent of the intake at elite universities (Hazell 2018). Russell Group institutions have to reconcile their responsibility to work for inclusion with admission requirements that reflect their essentially selective place in a hyper-competitive, marketised HE regime (Collini 2011). As we have seen, HE admissions widened to include almost 50 per cent of young people by 2017 but participation rates amongst advantaged and disadvantaged undergraduates did not become more equal (Smith 2011). The circle can be fudged but not squared.

The Socio-Economic Duty

The Equality Act 2010 represented a major step forward for the equal rights movement and recognised the legal rights of a wider range of disadvantaged groups. A new 'socio-economic duty' required public bodies to operate in ways that reduce socio-economic disadvantage. The duty was supposed to close the gap between rich and poor. Hepple (2010: 21) argues that 'the streamlining and broadening of the public sector equality duty and the provisions on permissible positive action are at the core of the new approach to *transformative* equality.' Ten years later, however, this interpretation is unconvincing. The socio-economic duty has not been implemented, legal aid has been dramatically reduced, and as the Brexit debate has unfolded, the political climate has become less sympathetic to human rights. As the EHRC reports

> ...these rights mean nothing without a level playing field between ordinary people and organisations able to afford professional legal representation.

[1]POLAR (Participation of Local Areas) is a measure commonly used by universities to discover the extent to which an applicant may have faced disadvantage. The country is divided into a large number of 'local areas'; the government compiles statistics on how many young people in each area typically go to university. The 20% of areas with the lowest participation rates are designated quintile 1; the top 20% are quintile 5.

> And if the rights mean nothing, they do not operate as the powerful deterrent we need to effect lasting change and build a fairer society. (EHRC 2019: 3)

Another problem has become increasingly clear. The protected characteristics do not include social class and fail to recognise adequately the intersectionality of multiple identities and the synergy between the protected characteristics. Inequality, social class and injustice are as powerful as ever but the battle for human rights and diversity continue in unpromising circumstances. Our fifth proposition is therefore justified in pointing out the weaknesses of the Equality Act and confirms the further policy development needed to enhance the drive for equal rights.

The Future

Marsden chemistry has encountered barriers in working for equal rights and opportunities that seem beyond the scope of a single department, university or even HE itself. But victories have been won nevertheless. Women are admitted in equal numbers at undergraduate level and significant changes have been made to facilitate female careers, even if the levels of investment are inadequate to bring about a far-reaching transformation. Serious support is provided and appreciated by LGBTQI+ students and the department's stance and culture are instrumental in attracting candidates who feel vulnerable. There is an inclusive culture and discrimination is not acceptable. But the drive to increase diversity and extend the representation of disadvantaged students will continue to falter without national recognition that an improved and more vigorous approach to equal opportunities is required. Individualist discourses with their emphasis on merit and hard work overcoming disadvantage are not enough. The Civil Service provides an example of how a large national employer can achieve encouraging results by choosing to take the socio-economic duty seriously.

The Bridge Group (2016: 1) documents the Civil Service's work to promote socio-economic diversity in Fast Stream recruitment. Selection processes include the collection and publication of socio-economic background (SEB) data, the removal of candidate screening criteria (e.g. UCAS points and university attended), the targeting of marketing activity based on campus diversity, and an internship programme exclusively for under-represented groups. As a result, the Fast Stream intake from lower socio-economic backgrounds (SEB) has doubled, although it remains unrepresentative of the population at large (Friedman and Laurison 2020). BAME and disabled candidates, however, are much better represented. A senior civil servant says 'the Fast Stream achieved the highest ever diversity success rates in the majority of categories in 2019. Representation in Fast Stream appointments is now above our benchmarks for gender and LGBO' (Toogood 2019: unpaged).

Although the Bridge Group (2016) acknowledges more needs to be done, especially about the under-representation of lower SEB candidates, this is an important example. It shows the need for a systematic, national approach to ensuring equal opportunities in recruitment. Through Athena Swan, HE is attempting to match the Bridge Group's initiative but lacks the system-wide control available to Civil Service managers. Individual departments and universities cannot resolve recruitment issues that arise from a league table-driven market system. The selective admissions policies adopted by elite universities attract confident middle class students with excellent track records at school; they inevitably discourage less confident, lower SEB applicants who doubt their prospects of success. The HE system resembles the pattern of secondary education before comprehensive reorganisation; and has similar social-sorting effects.

The current context is discouraging but there are grounds for optimism about equal opportunities. 'Fairness' is a core British value and the Bridge Group and Athena Swan have pioneered significant progress. *Aspires 2* (Archer et al. 2020) has identified changes in practice with the potential to improve access and build young people's science capital at school level. But we also need HE to become less selective; we need to value a much wider range of skills and abilities; we need Athena Swan principles to be backed up with teeth; and above all, we need governments ready to implement the socio-economic duty. They should follow the example of the Fast Stream, where recruitment aims to produce a workforce that is more representative of our wider society. Our study argues that current policies are based on a misleading impression of how aspirations are formed and how young people's decisions are made. The conclusion is clear. We should invest more in supporting disadvantaged families, and must also ensure that equal rights legislation is fully effective in outlawing discrimination on grounds of class, gender and ethnicity.

References

Archer, L., DeWitt, J., Osborne, J., Dillon, J., Willis, B., & Wong, B. (2012). Science Aspirations, Capital, and Family Habitus: How Families Shape Children's Engagement and Identification with Science. *American Educational Research Journal, 49*(5), 881–908.

Archer, L., Moote, J., MacLeod, E., Francis, B., & DeWitt, J. (2020). *ASPIRES 2: Young People's Science and Career Aspirations, Age 10–19*. London: UCL Institute of Education.

Ball, S., Maguire, M., & Macrae, S. (2000). *Choice, Pathways and Transitions Post-16: New Youth, New Economies in the Global City*. London: Routledge Falmer.

Barker, B. (2010). *The Pendulum Swings: Transforming School Reform*. Stoke-on-Trent: Trentham Books.

Bernardi, F., & Ballarino, G. (Eds.). (2016). *Education, Occupation and Social Origin: A Comparative Analysis of the Transmission of Socio-Economic Inequalities*. Cheltenham: Elgar.

Bourdieu, P. (1986). The Forms of Capital. In J. Richardson (Ed.), *Handbook of Theory of Research for the Sociology of Education* (pp. 241–258). New York: Greenwood Press. Translated by R. Nice.

Bourdieu, P., & Passeron, J.-C. (1977). *Reproduction in Education, Society and Culture*. London: Sage.

Bridge Group, CEM. (2016). *Socio-economic Diversity in the Fast Stream*. London: Centre for Evaluation and Monitoring.

Brown, P. (2013). Education, Opportunity and The Prospects for Social Mobility. *British Journal of Sociology of Education, 34*(5–6), 678–700.

Bukodi, E., & Goldthorpe, J. (2018). *Social Mobility and Education in Britain: Research, Politics and Policy*. Cambridge: Cambridge University Press.

Chilisa, B. (2006). Sex Education: Subjugated Discourses and Adolescents' Voice. In C. Skelton, B. Francis, & L. Smulyan (Eds.), *The Sage Handbook of Gender and Education* (pp. 249–261). London: Sage Publications.

Collini, S. (2011). From Robbins to McKinsey. *London Review of Books, 33*(16), 9–14.

Elgot, J. (2018, August 15). Government Accused of 'Total Failure' to Widen Elite University Access. *The Guardian*. Retrieved August 15, 2018, from https://www.theguardian.com/education/2018/aug/15/government-accused-of-total-failure-to-widen-elite-university-access.

Equality and Human Rights Commission (EHRC). (2019). *Access to Legal Aid for Discrimination Cases*. EHRC. Retrieved December 31, 2019, from https://www.equalityhumanrights.com/sites/default/files/access-to-legal-aid-for-discrimination-cases-our-legal-aid-inquiry.pdf.

Friedman, S., & Laurison, D. (2020). *The Class Ceiling: Why It Pays to Be Privileged*. Bristol: Policy Press.

Gladwell, M. (2008). *Outliers: The Story of Success*. New York: Little Brown.

Goldthorpe, J. (2007). 'Cultural Capital': Some Critical Observations. *Sociologica, September-October, 2*, 1–23.

Goldthorpe, J. (2014). The Role of Education in Intergenerational Social Mobility: Problems from Empirical Research in Sociology and Some Theoretical Pointers from Economics. *Rationality and Society, 26*(3), 265–289.

Greenfield, S., Peters, J., Lane, N., Rees, T., & Samuels, G. (2002). *Set Fair, A Report on Women in Science, Engineering, and Technology*. London: Department of Trade and Industry.

Hall, R. (2017, September 6). Les Ebdon: 'We Need a Step Change on Widening Access to University'. *The Guardian*. Retrieved April 3, 2019, from https://www.theguardian.com/higher-education-network/2017/sep/06/les-ebdon-university-widening-access-office-fair-access-students.

Hazell, W. (2018, January 25). New Figures Reveal Dearth of Poor Students at Russell Group Universities. *Times Educational Supplement*. Retrieved December 31, 2019, from https://www.tes.com/news/new-figures-reveal-dearth-poor-students-russell-group-universities.

Health and Social Care Committee (HSCC) SCC. (2019). First 1000 Days of Life. Retrieved December 31, 2019, from https://publications.parliament.uk/pa/cm201719/cmselect/cmhealth/1496/1496.pdf.

Hepple, B. (2010). The New Single Equality Act in Britain. *The Equal Rights Review, 5*, 11–24.

Hoskins, K., & Barker, B. (2019). Social Mobility: The Potential of a Genealogical Approach. *British Educational Research Journal, 45*(2), 238–253.

Mandler, P. (2017). Educating the Nation: IV. Subject Choice. *Transactions of the Royal Historical Society, 27*, 1–27.

McKie, R. (2019, August 31). Fears for Britain's Standing in World of Science as Students Shun Chemistry Degrees. *The Guardian*. Retrieved November 25, 2019, from https://www.theguardian.com/education/2019/aug/31/students-shun-chemistry-degrees-university-applications-fall.

O'Hara, M. (2014). *Austerity Bites: A Journey to the Sharp End of Cuts in the UK*. Bristol: Polity Press.

Pearce, R. (2017). Certifying Equality: Critical reflections on Athena Swan and Equality Accreditation, Centre for the Study of Women & Gender, University of Warwick. Retrieved June 3, 2018, from https://warwick.ac.uk/fac/soc/sociology/research/centres/gender/calendar/certifying_equality_a/certifying_equality_-_critical_reflection_on_athena_swan.pdf.

Reay, D. (2017). *Miseducation: Inequality, Education and the Working Classes*. Bristol: Policy Press.

Smith, E. (2011). Women into Science and Engineering? Gendered Participation in Higher Education STEM Subjects. *British Educational Research Journal, 37*(6), 993–1014.

Social Mobility and Child Poverty Commission (SMCPC). (2014). State of the Nation 2014: Social Mobility and Child Poverty in Great Britain, Report Summary. Retrieved July 16, 2017, from https://assets.publishing.service.gov.uk/government/uploads/system/uploads/attachment_data/file/367461/State_of_the_Nation_-_summary_document.pdf.

Social Mobility Commission (SMC). (2016). State of the Nation 2016: Social Mobility in Great Britain. Retrieved April 11, 2018, from www.assets.publishing.service.gov.uk/government/uploads/system/uploads/attachment_data/file/569412/Social_Mobility_2016_Summary_final.pdf.

Social Mobility Commission (SMC). (2019). State of the Nation 2018–19: Social Mobility in Great Britain. Retrieved December 12, 2019, from https://assets.publishing.service.gov.uk/government/uploads/system/uploads/attachment_data/file/798687/SMC_State_of_Nation_2018-19_Summary.pdf.

Swartz, B. (1997). *Culture and Power: The Sociology of Pierre Bourdieu*. Chicago, IL: University of Chicago Press.

Targeted Initiative on Science and Mathematics Education (TISME). (2013). What Influences Participation in Science and Mathematics? Retrieved August 28, 2015, from https://www.kcl.ac.uk/sspp/departments/education/research/aspires/TISME-briefing-paper-March-2013.pdf.

Thompson, P. (1997). Women, Men, and Transgenerational Family Influences in Social Mobility. In D. Bertaux & P. Thompson (Eds.), *Pathways to Social Class: A Qualitative Approach to Social Mobility* (pp. 32–61). Oxford: Clarendon Press.

Toogood, J. (2019). Civil Service: Inclusion in the Fast Stream, Blog. Retrieved January 6, 2020, from https://civilservice.blog.gov.uk/2019/09/27/inclusion-in-the-fast-stream/.

Wilkinson, R., & Pickett, K. (2009). *The Spirit Level: Why Equality Is Better for Everyone*. Harmondsworth: Penguin Books.

Appendix

K. Hoskins, B. Barker, *STEM, Social Mobility and Equality*,
https://doi.org/10.1007/978-3-030-49216-8

Occupations Table

Student	*Maternal Grandfather*	*Maternal Grandmother*	*Mother*	*Paternal Grandfather*	*Paternal Grandmother*	*Female*	*Self-Rate*
Ainsley	**Electrician** WC	Housewife WC	OU graduate BAE **Engineer with physics/maths background** UWC or LMC	**Electrician** WC	Office secretary UWC or LMC	OU graduate BAE **Engineer with physics/maths background** Uni IT technician UWC or LMC	UWC or LMC First formal university educated
Anna	Manual labourer & factory worker WC	Clerical and shop work WC	Teacher, SENCO, SEN assessor UWC	**Toolmaker, machinery designer** UWC	Estate agent, child minder UWC	**Mechanical engineer** UWC	UWC
Bella	Teacher Mining background MC	Teacher Mining background MC	**Alternative Therapist (science degree)** MC	Health service manager UWC	Secretary UWC	Client relationship executive (Global IT) **(science degree)** MC	MC
Carrie	Miner and then coach driver WC	Cruise ship & chef, widely travelled (from Welsh valleys) WC	Teaching assistant MC	General labourer with farming background MC	**Nurse** (but medical condition) MC	Director of landscape gardening company MC	MC
Daniella	**Lecturer/ researcher in chemistry** MC	**University demonstrator (chemistry)** MC	Entrepreneur specialising in property investment, recent degree UMC	High ranking army officer, construction company and farm MC	**Psychology expert witness/lecturer** MC	Step: Barrister UMC Bio: Company director making car lifts MC	UMC (but NB translation from Eastern European context)

Student	M.GF	M.GM	M	P.GF	P.GM	F	Self-Rate
David	Head of synthetic silk factory **accounts department** UMC	Worked in silk **factory accounts** and housewife UMC	Language and vocational subjects teacher (Lithuania) LMC	Tractor driver on collective farm LWC	Housewife LWC	**Metallurgist (degree) researcher.** Now long-haul truck driver because factories closed LMC	MC
Gabriella	Estranged (no influence) WC	**Accounts worker**, secretary, kitchens, bars, single parent WC	**Auditor, accountant**, business owner and kitchen bar staff WC	Landowner & worker (olives & grapes), tractor part production WC (Migrants from Italy)	Agricultural worker and manager (grapes) WC (Migrants from Italy)	Tailor's assistant, National Service, fishmonger & head chef—chef since 15 migrating for work WC	W/ LMC
Gail	Hardware store owner and **electrical engineer** WC	Hairdresser WC	Self-employed owner of storage business WC	Crane driver WC	**Nurse** WC	Self-employed owner of storage business WC	W/ LMC
Joel	**Electrical engineer** MC	Lecturer and teacher in education and sociology MC	**R&D at Unilever; secondary school teacher, has PhD**, ibid her brother MC	Civil engineer; taught special needs children WC	Uni lecturer in art and graphics WC	**IT consultant** for HP MC	MC
George	Insurance sales, publican, postmaster WC/ LMC	**Head of maths** after office roles, bar staff WC/ LMC	**Nurse, senior lecturer in nursing** and higher education coordinator WC	Self-employed butcher, police special in war WC	Cobbler in shoe shop, shop assistant and community carer WC	**Civil engineer** with British Waterways, sales manager, site/plant manager MC	UWC/ LMC

(continued)

(continued)

Student	M.GF	M.GM	M	P.GF	P.GM	F	Self-Rate
Kyle	**Chemistry teacher**, education adviser and vocational skills trainer for a government department MC	Middle school teacher MC	**Environmental health officer** MC →	Army lieutenant, secondary teacher & college university lecturer MC	Secondary teacher; special needs, head of year MC	Pharmaceutical manufacturing for GSK, from factory director to **analytical development chemist** MC ↑	MC

Key: *WC* Working Class; *UWC* Upper Working Class; *LMC* Lower Middle Class; *UMC* Upper Middle Class

Upward Mobility in generation 2: ↑ Comparable with parents in generation 2: → . STEM background employment: in bold font

References

Adams, R. (2017, September 28). Almost Half of All Young People in England Go on to Higher Education. *The Guardian*. Retrieved May 1, 2019, from https://www.theguardian.com/education/2017/sep/28/almost-half-of-all-young-people-in-england-go-on-to-higher-education.

Adonis, A. (2013, January 4). How Labour Can Learn from Roy Jenkins. *The Guardian*. Retrieved May 1, 2019, from https://www.theguardian.com/commentisfree/2013/jan/04/how-labour-can-learn-roy-jenkins.

Advanced Resource Managers (ARM). (2016). *The Truth Behind The STEM Skill Shortage: A Serious Mismatch Between Hiring Company Perceptions and Job Seeking Reality*. ARM White Paper, June. Retrieved June 3, 2019, from https://www.arm.co.uk/media/1645/arm-the-truth-behind-the-stem-skill-shortage.pdf.

AdvanceHE. (2019). AdvanceHE's Athena SWAN Charter. Retrieved February 1, 2020, from https://www.ecu.ac.uk/equality-charters/athena-swan/.

Almeida, R., Arbelaez, J., Honorati, M., Kuddo, A., Lohmann, T., Ovadiya, M., Pop, L., Puerta, M., & Weber, M. (2012). *Improving Access to Jobs and Earnings Opportunities: The Role of Activation and Graduation Policies in Developing Countries*. Social Protection and Labor Discussion Paper No. 1204. Retrieved June 3, 2019, from http://siteresources.worldbank.org/SOCIALPROTECTION/Resources/SP-Discussion-papers/430578-1331508552354/1204.pdf.

Andrews, J., Robinson, D., & Hutchinson, J. (2017). *Closing the Gap? Trends in Educational Attainment and Disadvantage*. Education Policy Institute. Retrieved June 3, 2019, from https://epi.org.uk/wp-content/uploads/2017/08/Closing-the-Gap_EPI-.pdf.

K. Hoskins, B. Barker, *STEM, Social Mobility and Equality*,
https://doi.org/10.1007/978-3-030-49216-8

Archer, L., & Tomei, A. (2013). *What Influences Participation in Science and Mathematics? A Briefing Paper from the Targeted Initiative on Science and Mathematics Education (TISME)*. King's College London.

Archer, L., DeWitt, J., Osborne, J., Dillon, J., Willis, B., & Wong, B. (2012). Science Aspirations, Capital, and Family Habitus: How Families Shape Children's Engagement and Identification with Science. *American Educational Research Journal, 49*(5), 881–908.

Archer, L., Osborne, J., DeWitt, J., Dillon, J., Wong, B., & Willis, B. (2013). Aspires: Young People's Science and Career Aspirations, Age 10–14, Department of Education and Professional Studies, King's College, London. Funded by the ESRC, Research Grant Number RES 179-25-0008, as Part of the Targeted Initiative on Science and Mathematics Education (TISME). Retrieved August 28, 2015, from https://www.kcl.ac.uk/ecs/research/aspires/aspires-final-report-december-2013.pdf.

Archer, L., Dawson, E., DeWitt, J., Seakins, A., & Wong, B. (2015). "Science Capital": A Conceptual, Methodological, and Empirical Argument for Extending Bourdieusian Notions of Capital Beyond the Arts. *Journal of Research in Science Teaching, 52*(7), 922–948.

Archer, L., Nomikou, E., Mau, A., King, H., Godec, S., DeWitt, J., & Dawson, E. (2018). Can the Subaltern 'Speak' Science? An Intersectional Analysis of Performances of 'Talking Science Through Muscular Intellect' by 'Subaltern' Students in UK Urban Secondary Science Classrooms. *Cultural Studies of Science Education*. https://doi.org/10.1007/s11422-018-9870-4.

Archer, L., Moote, J., MacLeod, E., Francis, B., & DeWitt, J. (2020). *ASPIRES 2: Young People's Science and Career Aspirations, Age 10–19*. London: UCL Institute of Education.

Ashenden, A., & Parsons, V. (2019, June 13). 15 Rights LGBT People in the UK Still Don't Have. *Pink News*. Retrieved June 3, 2019, from https://www.pinknews.co.uk/2019/06/13/15-rights-lgbt-uk-still-dont-have-pride/.

Atkinson, A. (2015). *Inequality. What Can Be Done?* Cambridge, MA: Harvard University Press.

Attar, N. (2012). *Susan Greenfield on Women in Science: Not Much Has Improved in 10 Years*. BMC/Springer, Blog Network. Retrieved September 1, 2018, from https://blogs.biomedcentral.com/on-biology/2012/03/08/susan-greenfield-on-women-in-science-not-much-has-improved-in-10-years/.

Avendano, L., Renteria, J., Kwon, S., & Hamdan, K. (2019). Bringing Equity to Underserved Communities Through STEM Education: Implications for Leadership Development. *Journal of Educational Administration and History, 51*(1), 66–82.

Bagilhole, B. (2007). Challenging Women in the Male Academy: Think About Draining the Swamp. In P. Cotterill, S. Jackson, & G. Letherby (Eds.),

Challenges and Negotiations for Women in Higher Education (pp. 21–32). Dordrecht, The Netherlands: Springer.

Ball, S. (2003). *Class Strategies and the Education Market: The Middle Class and Social Advantage*. London: Routledge Falmer.

Ball, S., Maguire, M., & Macrae, S. (2000). *Choice, Pathways and Transitions Post-16: New Youth, New Economies in the Global City*. London: Routledge Falmer.

Barber, M. (2007). *Instruction to Deliver*. London: Politico's Publishing.

Barker, B. (2010). *The Pendulum Swings: Transforming School Reform*. Stoke-on-Trent: Trentham Books.

Barker, B., & Hoskins, K. (2015). Can High-Performing Academies Overcome Family Background and Improve Social Mobility? *British Journal of Sociology of Education, 38*(2), 221–240.

Barnett, C. (1986). *The Audit of War*. Basingstoke and Oxford: Macmillan.

Barr, B., Kinderman, P., & Whitehead, M. (2015). Trends in Mental Health Inequalities in England During a Period of Recession, Austerity and Welfare Reform 2004 to 2013. *Social Science and Medicine, 147*, 324–331.

Bell, D. (2015). Perceptions of STEM, Within the Context of Teaching Design and Technology in Secondary Schools; A Phenomenographic Study. Edge Hill University. Retrieved May 1, 2019, from https://repository.edgehill.ac.uk/6331/1/.

Belmi, P., Neale, M. A., Reiff, D., & Ulfe, R. (2020). The Social Advantage of Miscalibrated Individuals: The Relationship Between Social Class and Overconfidence and Its Implications for Class-Based Inequality. *Journal of Personality and Social Psychology, 118*(2), 254–282.

Benn, C., & Chitty, C. (1996). *Thirty Years On: Is Comprehensive Education Alive and Well Or Struggling to Survive?* London: David Fulton Publishers, Ltd.

Berger, P., & Luckmann, T. (1966). *The Social Construction of Reality: A Treatise in the Sociology of Knowledge*. New York: Random House.

Bernardi, F., & Ballarino, G. (Eds.). (2016). *Education, Occupation and Social Origin: A Comparative Analysis of the Transmission of Socio-Economic Inequalities*. Cheltenham: Elgar.

Bertaux, D. (1995). Social Genealogies Commented on and Compared: An Instrument for Observing Social Mobility Processes in the 'Longue Durée'. *Current Sociology, 43*(2), 69–88.

Bertaux, D., & Bertaux-Wiame, I. (1997). Heritage and Its Lineage: A Case History of Transmission and Social Mobility Over Five Generations. In D. Bertaux & P. Thompson (Eds.), *Pathways to Social Class: A Qualitative Approach to Social Mobility* (pp. 63–97). Oxford: Clarendon Press.

Bertaux, D., & Thompson, P. (Eds.). (1997). *Pathways to Social Class: A Qualitative Approach to Social Mobility*. Oxford: Clarendon Press.

Bhopal, K. (2018). *White Privilege: The Myth of a Post-racial Society*. Bristol: Policy Press.

Blair, T. (2000, March 7). Tony Blair's Full Speech, 'The e-generation is with us'. *The Guardian*. Retrieved July 20, 2016, from https://www.theguardian.com/uk/2000/mar/07/tonyblair.

Blair, T. (2001, May 23). Full Text of Tony Blair's Speech on Education. *The Guardian*. Retrieved July 20, 2016, from https://www.theguardian.com/politics/2001/may/23/labour.tonyblair.

Blair, T. (2002, June 10). Full Text of Tony Blair's Speech on Welfare Reform. *The Guardian*. Retrieved July 20, 2016, from http://www.guardian.co.uk/society/2002/jun/10/socialexclusion.politics.

Blanden, J., & Machin, S. (2013). Educational Inequality and the Expansion of United Kingdom Higher Education. *Scottish Journal of Political Economy, 60*, 5.

Blanden, J., Gregg, P., & Machin, S. (2005a). *Intergenerational Mobility in Europe and North America*. London: Centre for Economic Performance, London School of Economics.

Blanden, J., Gregg, P., & Machin, S. (2005b). Social Mobility in Britain: Low and Falling. *CentrePiece, 10*, 18–20.

Bloome, D. (2014). Racial Inequality Trends and the Intergenerational Persistence of Income and Family Structure. *American Sociological Review, 79*(6), 1196–1225.

Bolton, P. (2012). *Education: Historical Statistics*. House of Commons Library. Retrieved July 20, 2016, from https://dera.ioe.ac.uk//22771/.

Bourdieu, P. (1977). *Outline of a Theory of Practice*. Cambridge: Cambridge University Press.

Bourdieu, P. (1979). *Algeria 1960: The Disenchantment of the World, the Sense of Honour, the Kabyle House Or the World Reversed*. Cambridge: Cambridge University Press.

Bourdieu, P. (1984). *Distinction*. London: Routledge and Kegan Paul.

Bourdieu, P. (1986). The Forms of Capital. In J. Richardson (Ed.), *Handbook of Theory of Research for the Sociology of Education* (pp. 241–258). New York: Greenwood Press. Translated by R. Nice.

Bourdieu, P. (1990). *Sociology in Question*. Cambridge: Polity Press.

Bourdieu, P., & Passeron, J.-C. (1977). *Reproduction in Education, Society and Culture*. London: Sage.

Brah, A., & Phoenix, A. (2004). Ain't I a Woman? Revisiting Intersectionality. *Journal of International Women's Studies, 5*(3), 75–86.

Breiner, J., Harkness, S., Johnson, C., & Koehler, C. (2012). What Is STEM? A Discussion about Conceptions of STEM in Education and Partnerships. *School Science and Mathematics, 112*(1), 3–11.

Bridge Group, CEM. (2016). *Socio-economic Diversity in the Fast Stream*. London: Centre for Evaluation and Monitoring.

British Broadcasting Corporation (BBC) News. (2000). Chancellor Attacks Oxford Admissions. Retrieved March 23, 2018, from http://news.bbc.co.uk/1/hi/education/764141.stm.

British Broadcasting Corporation (BBC) News. (2013). The Great British Class Calculator: What Class Are You? Retrieved June 3, 2018, from www.bbc.co.uk/news/magazine-22000973.

British Broadcasting Corporation (BBC) News Business. (2018, April 19). Apprenticeships Down Amid 'Frustration'. Retrieved May 15, 2019, from https://www.bbc.co.uk/news/business-43809679.

British Educational Research Association (BERA). (2018). Ethical Guidelines for Educational Research. Retrieved December 31, 2018, from http://www.bera.ac.uk/content/ethical-guidelines.

British Sociological Association (BSA). (2002). Statement of Ethical Practice for the British Sociological Association. Retrieved December 31, 2019, from www.britsoc.ac.uk.

Broecke, S., & Hamed, J. (2008). *Gender Gaps in Higher Education Participation: An Analysis of the Relationship Between Prior Attainment and Young Participation by Gender, Socio-Economic Class and Ethnicity*. DIUS Research Report 08 14, Department for Innovation, Universities and Skills. Retrieved July 15, 2015, from http://webarchive.nationalarchives.gov.uk/+/http:/www.dius.gov.uk/research/documents/DIUS-RR-08-14.pdf.

Brown, D. (2001). The Social Sources of Educational Credentialism: Status Cultures, Labor Markets, and Organizations. *Sociology of Education, 74* (Extra Issue: Current of Thought: Sociology of Education at the Dawn of the 21st Century), 19–34.

Brown, P. (2013). Education, Opportunity and The Prospects for Social Mobility. *British Journal of Sociology of Education, 34*(5–6), 678–700.

Buckingham, J. (2019). Athena Swan Review – Update from the Steering Group, AdvanceHE. Retrieved January 1, 2020, from https://www.advance-he.ac.uk/news-and-views/Athena%20SWAN%20Review%20-%20update%20from%20the%20Steering%20Group.

Bukodi, E., & Goldthorpe, J. (2018). *Social Mobility and Education in Britain: Research, Politics and Policy*. Cambridge: Cambridge University Press.

Bukodi, E., Erikson, R., & Goldthorpe, J. (2014). The Effects of Social Origins and Cognitive Ability on Educational Attainment: Evidence from Britain and Sweden. *Acta Sociologica, 57*(4), 293–310.

Bunting, M. (2007, January 24). The Class Debate Demands to Be Heard. *The Guardian*. Retrieved December 31, 2019, from https://www.theguardian.com/commentisfree/2007/jan/24/society.comment.

Burr, V. (2003). *Social Constructionism* (2nd ed.). East Sussex: Routledge.

Burr, V. (2015). *Social Constructionism* (3rd ed.). East Sussex: Routledge.

Business in the Community (BITC). (2012). *Women and Work: The Facts.* Fact Sheet. Retrieved July 20, 2016, from http://opportunitynow.bitc.org.uk/research-insight/WomenWorkFactsheet.

Butler, P. (2016, June 29). Most Britons Regard Themselves as Working Class, Survey Finds. *The Guardian.* Retrieved April 24, 2018, from https://www.theguardian.com/society/2016/jun/29/most-brits-regard-themselves-as-working-class-survey-finds.

Byrne, A., Shlomo, N., & Chandola, T. (2017). Multilevel Modelling Approach to Analysing Life Course Socioeconomic Status and Compensating for Missingness. Retrieved July 12, 2019, from http://hummedia.manchester.ac.uk/institutes/cmist/archive-publications/working-papers/2017/MLM%20approach%20to%20analysing%20life%20course%20SES%20and%20compensating%20for%20missingness.pdf.

Callender, C., & Mason, G. (2017). Does Student Loan Debt Deter Higher Education Participation? New Evidence from England. *Annals of American Political and Social Science, 671*(1), 20–48.

Canetto, S., Trott, C., Winterrowd, E., Haruyama, D., & Johnson, A. (2017). Challenges to the Choice Discourse: Women's Views of Their Family and Academic-Science Career Options and Constraints. *Journal of Feminist Family Therapy, 29*(1–2), 4–27.

Cassidy, R., Cattan, S., Crawford, C., & Dytham, S. (2019). How Can We Increase Girls' Uptake of Maths and Physics A Level? Institute for Fiscal Studies. Retrieved December 31, 2019, from https://www.ifs.org.uk/uploads/R149%20with%20cover.1.pdf.

Chan, T. W., & Boliver, V. (2013). The Grandparents Effect in Social Mobility: Evidence from British Birth Cohort Studies. *American Sociological Review, 78*(4), 662–678.

Chappell, A., McHugh, E., & Wainwright, E. (2018). *Successful Students: Exploring the Factors That Encourage and Enable Students from a Widening Participation Background to Stay the Course (Report).* Uxbridge: Brunel University London.

Chilisa, B. (2006). Sex Education: Subjugated Discourses and Adolescents' Voice. In C. Skelton, B. Francis, & L. Smulyan (Eds.), *The Sage Handbook of Gender and Education* (pp. 249–261). London: Sage Publications.

Collini, S. (2011). From Robbins to McKinsey. *London Review of Books, 33*(16), 9–14.

Collins, R. (1981). Crises and Declines in Credentialing Systems. In R. Collins (Ed.), *Sociology Since Mid-Century* (pp. 191–215). New York: Academic Press.

Collins, P., & Bilge, S. (2016). *Intersectionality.* Cambridge: Polity Press.

Cook, C. (2012, February 22). The Social Mobility Challenge for School Reformers. *FT Data.* Retrieved March 11, 2012, from http://blogs.ft.com/

ftdata/2012/02/22/social-mobility-and-schools/#axzz1o06SJkBJ (Registration Required).

Cornwell, J. (2004). *Hitler's Scientists: Science, War and the Devil's Pact*. London: Penguin Books.

Cotterill, P., Jackson, S., & Letherby, G. (Eds.). (2007). *Challenges and Negotiations for Women in Higher Education*. The Netherlands: Springer.

Crenna-Jennings, W. (2018). *Education in England: Annual Report*. Education Policy Institute. Retrieved April 11, 2019, from https://epi.org.uk/wp-content/uploads/2018/07/EPI-Annual-Report-2018-Lit-review.pdf.

Crozier, G., Reay, D., James, D., Jamieson, F., Beedell, P., Hollingworth, S., & Williams, K. (2008). White Middle-Class Parents, Identities, Educational Choice and the Urban Comprehensive School: Dilemmas, Ambivalence and Moral Ambiguity. *British Journal of Sociology of Education, 29*(2), 61–272.

Darton, D., & Strelitz, J. (Eds.). (2004). *Tackling UK Poverty and Disadvantage in the Twenty-First Century: An Exploration of the Issues*. York: Joseph Rowntree Foundation.

David, M., & Woodward, D. (1998). *Negotiating the Glass Ceiling: Careers of Senior Women in the Academic World*. London: Routledge.

Department for Education (DfE). (2010a). *The Importance of Teaching: The Schools White Paper*. London: The Stationery Office.

Department for Education (DfE). (2010b). *The Importance of Teaching White Paper Equalities Impact Assessment (CM-7980)*. London: The Stationery Office.

Department for Education (DfE). (2018). *Widening Participation in Higher Education, England, 2016/17 Age Cohort – Official Statistics*. GOV.UK.

Department for Education and Skills (DFES). (2004). Report on the Science, Technology, Engineering and Maths (STEM) Mapping Review, Volume 1, Main Report, Higher Education Directorate.

Dex, S., Bukodi, E., & Joshi, H. (2009, December 8). *The Ups and Downs of Men's and Women's Careers*. Institute of Education, GeNet Seminar, London. Retrieved July 14, 2015, from http://www.cls.ioe.ac.uk.

DfE. (2017). *Unlocking Talent, Fulfilling Potential: A Plan for Improving Social Mobility Through Education*, CM 9541.

Disability Rights UK. (2012). Campaigns. Retrieved July 1, 2015, from https://www.disabilityrightsuk.org/policy-campaigns/campaigns.

Dorling, D. (2014a). *Inequality and the 1%*. London: Verso Books.

Dorling, D. (2014b). Thinking About Class. *Sociology, 48*(3), 452–462.

Dryden, S. (2018). *A Short History of LGBT Rights in the UK*. British Library. Retrieved December 31, 2019, from https://www.bl.uk/lgbtq-histories/articles/a-short-history-of-lgbt-rights-in-the-uk.

Dyhouse, C. (2002). Graduates, Mothers and Graduate Mothers: Family Investment in Higher Education in Twentieth-Century England. *Gender and Education, 14*(4), 325–333.

Elgot, J. (2018, August 15). Government Accused of 'Total Failure' to Widen Elite University Access. *The Guardian*. Retrieved August 15, 2018, from https://www.theguardian.com/education/2018/aug/15/government-accused-of-total-failure-to-widen-elite-university-access.

Equality and Human Rights Commission (EHRC). (2016). Healing a Divided Britain: The Need for a Comprehensive Race Equality Strategy. Retrieved December 31, 2019, from https://www.equalityhumanrights.com/en/publication-download/healing-divided-britain-need-comprehensive-race-equality-strategy.

Equality and Human Rights Commission (EHRC). (2019). *Access to Legal Aid for Discrimination Cases*. EHRC. Retrieved December 31, 2019, from https://www.equalityhumanrights.com/sites/default/files/access-to-legal-aid-for-discrimination-cases-our-legal-aid-inquiry.pdf.

Equality Challenge Unit (ECU). (2010). Advancing LGB Equality: Improving the Experience of Lesbian, Gay and Bisexual Staff and Students in Higher Education. Retrieved June 4, 2019, from www.ecu.ac.uk/publications/advancing-lgb-equality.

Erikson, R., & Goldthorpe, J. (2010). Has Social Mobility in Britain Decreased? Reconciling Divergent Findings on Income and Class Mobility. *The British Journal of Sociology, 61*(2), 211–230.

Erola, J., & Moisio, P. (2007). Social Mobility Over Three Generations in Finland 1950–2000. *European Sociological Review, 23*, 169–183.

Ferguson, D. (2017, November 21). Working Class Children Get Less of Everything in Education – Including Respect. *The Guardian*. Retrieved November 21, 2017, from https://www.theguardian.com/education/2017/nov/21/english-class-system-shaped-in-schools.

Fortin, N., Bell, B., & Bohm, M. (2017). Top Earnings Inequality and the Gender Pay Gap: Canada, Sweden, and the United Kingdom. *Labour Economics, 47*, 107–123.

Fowler, S. (2019, January 10). Women in Stem: Critical to Innovation. *Global Policy Journal*: Wiley-Blackwell.

Francis, B. (2006). Troubling Trajectories: Gendered "Choices" and Pathways from School to Work. In C. Leathwood & B. Francis (Eds.), *Gender and Lifelong Learning: Critical Feminist Engagements* (pp. 57–69). London: Routledge.

Francis, R., Taylor, R., & Tereshchenko, A. (2019). *Reassessing 'Ability' Grouping: Improving Practice for Equity and Attainment*. London: Routledge.

Friedman, S., & Laurison, D. (2020). *The Class Ceiling: Why It Pays to Be Privileged*. Bristol: Policy Press.

Fumagalli, L. (2019). Short and Long Term Barriers to University Education in England. What Do Expectation Data Tell Us? ISER: University of Essex. Retrieved February 1, 2020, from https://www.iser.essex.ac.uk/files/projects/information-expectations-transitions-he/barriers-university-education-england.pdf.

Geertz, C. (1973). *The Interpretation of Cultures: Selected Essays*. New York: Basic Books.

Gewirtz, S. (2001). Cloning the Blairs: New Labour's Programme for the Resocialization of Working Class Parents. *Journal of Education Policy, 16*(4), 365–378.

Girl Scouts of the USA. (2012). *Generation Stem: What Girls Say About Science, Technology, Engineering, and Maths*. Report from the Girl Scout Research Institute. Retrieved October 25, 2018, from https://www.girlscouts.org/join/educators/generation_stem_full_report.pdf.

Gladwell, M. (2008). *Outliers: The Story of Success*. New York: Little Brown.

Goldin, C. (2014). Human Capital. In C. Diebolt & M. Haupert (Eds.), *Handbook of Cliometrics* (pp. 55–86). Heidelberg, Germany: Spring-Verlag.

Goldthorpe, J. (2007). 'Cultural Capital': Some Critical Observations. *Sociologica, September-October, 2*, 1–23.

Goldthorpe, J. (2014). The Role of Education in Intergenerational Social Mobility: Problems from Empirical Research in Sociology and Some Theoretical Pointers from Economics. *Rationality and Society, 26*(3), 265–289.

Goldthorpe, J., & Jackson, M. (2007). Intergenerational Class Mobility in Contemporary Britain: Political Concerns and Empirical Findings. *The British Journal of Sociology, 58*(4), 525–546.

Goldthorpe, J., & Mills, C. (2008). Trends in Intergenerational Class Mobility in Modern Britain: Evidence from National Surveys, 1972–2005. *National Institute Economic Review, 205*(1), 83–100.

Gorard, S. (2008). A Re-consideration of Rates of 'Social Mobility' in Britain: Or Why Research Impact Is Not Always a Good Thing. *The British Journal of Sociology of Education, 29*(3), 317–324.

Goulden, C. (2016). *There Can Be No Social Mobility Without More Affordable Housing*. Joseph Rowntree Foundation. Retrieved November 19, 2017, from https://www.jrf.org.uk/blog/there-can-be-no-social-mobility-without-more-affordable-housing.

Gove M. (2011). The Moral Purpose of School Reform, Address at the National College for School Leadership, Birmingham, 16 June, Unpaged. Retrieved March 30, 2018, from http://www.education.gov.uk/inthenews/speeches/a0077859/the-moral-purpose-ofschool-reform.

Gove, M. (2012). A Coalition for Good – How We Can All Work Together to Make Opportunity More Equal, Speech at Brighton College, 10 May, Unpaged. Retrieved March 30, 2018, from http://www.education.gov.uk/inthenews/speeches/a00208822/brightoncollege.

Government Equalities Office. (2018). *LGTB Action Plan: Improving the Lives of Lesbian, Gay, Bi-Sexual and Transgender People*. London: HMSO.

Greenfield, S., Peters, J., Lane, N., Rees, T., & Samuels, G. (2002). *Set Fair, A Report on Women in Science, Engineering, and Technology*. London: Department of Trade and Industry.

Haldane, J. (1938). *Science and Everyday Life*. Allahabad. First Indian Edition 1945: India Publishers.

Hall, R. (2017, September 6). Les Ebdon: 'We Need a Step Change on Widening Access to University'. *The Guardian*. Retrieved April 3, 2019, from https://www.theguardian.com/higher-education-network/2017/sep/06/les-ebdon-university-widening-access-office-fair-access-students.

Hazell, W. (2018, January 25). New Figures Reveal Dearth of Poor Students at Russell Group Universities. *Times Educational Supplement*. Retrieved December 31, 2019, from https://www.tes.com/news/new-figures-reveal-dearth-poor-students-russell-group-universities.

Health and Social Care Committee (HSCC) SCC. (2019). First 1000 Days of Life. Retrieved December 31, 2019, from https://publications.parliament.uk/pa/cm201719/cmselect/cmhealth/1496/1496.pdf.

Heath, A., & Payne, C. (1999). *Twentieth Century Trend in Social Mobility in Britain*. Working Paper No. 70, Centre for Research into Elections and Social Trends. Retrieved July 15, 2015, from http://www.crest.ox.ac.uk/papers/p70.pdf.

Hepple, B. (2010). The New Single Equality Act in Britain. *The Equal Rights Review, 5*, 11–24.

Higher Education Statistics Agency. (2013). Student Introduction 2012/13. Retrieved July 15, 2015, from https://www.hesa.ac.uk/intros/stuintro1213.

HM Government (HMG). (2009). *New Opportunities: Fair Chances for the Future*. Norwich: HMSO.

HM Government (HMG). (2010). *Unleashing Aspiration: The Government Response to the Final Report of the Panel on Fair Access to the Professions*. London: HMSO.

HM Government (HMG). (2011). *Opening Doors, Breaking Barriers: A Strategy for Social Mobility*. London: Cabinet Office.

HM Treasury (HMT). (2006). *Science and Innovation Investment Framework 2004–2014: Annual Report*. London: HM Treasury.

Hodkinson, P., Sparkes, A., & Hodkinson, H. (1996). *Triumphs and Tears: Young People, Markets and the Transition from School to Work*. London: David Fulton Publishers.

Holman, J., & Finegold, P. (2010). *Report to the Gatsby Charitable Foundation*. Warwick University. Retrieved April 15, 2018, from https://warwick.ac.uk/fac/soc/ier/ngrf/stem/movingon/research/500-stem_careers_review_nov_2010_holman.pdf.

Hoskins, K. (2012). *Women and Success: Professors in the UK Academy*. Stoke-on-Trent: Trentham Books.

Hoskins, K. (2017). *Youth Identities, Education and Employment Exploring Post-16 and Post-18 Opportunities, Access and Policy*. Basingstoke: Springer.

Hoskins, K., & Barker, B. (2014). *Education and Social Mobility: Dreams of Success*. London: Institute of Education Press.

Hoskins, K., & Barker, B. (2017). Aspirations and Young People's Constructions of Their Futures: Investigating Social Mobility and Social Reproduction. *British Journal of Educational Studies, 65*(1), 45–67.

Hoskins, K., & Barker, B. (2019). Social Mobility: The Potential of a Genealogical Approach. *British Educational Research Journal, 45*(2), 238–253.

Joannou, M., & Purvis, J. (Eds.). (1998). *The Women's Suffrage Movement: New Feminist Perspectives*. Manchester: Manchester University Press.

Jones, O. (2011). *Chavs: The Demonization of the Working Class*. London: Verso Books.

Kahn, O., & Sveinsson, K. (Eds.). (2015). *Race and Elections*. Runnymede Perspectives, Runnymede Trust. London: LSE.

Keck, T. (2009). Beyond Backlash: Assessing the Impact of Judicial Decisions on LGBT Rights. *Law & Society Review, 43*(1), 151–186.

Kings College London (KCL). (2019). TISME – Targeted Initiative on Science and Maths Education. Retrieved May 3, 2019, from https://www.kcl.ac.uk/ecs/research/research-centres/cppr/research/pastproj/tisme/index.

Kuenzi, J. (2008). Science, Technology, Engineering, and Mathematics (STEM) Education: Background, Federal Policy, and Legislative Action, Congressional Research Service Reports. 35. Retrieved October 2, 2019, from http://digitalcommons.unl.edu/crsdocs/35.

Lambert, P., Prandy, K., & Bottero, W. (2007). By Slow Degrees: Two Centuries of Social Reproduction and Mobility in Britain. *Sociological Research Online, 12*(1), 37–62.

Lareau, A. (2002). Invisible Inequality: Social Class and Childrearing in Black Families and White Families. *American Sociological Review, 67*(5), 747–776.

Lareau, A. (2003). *Unequal Childhoods: Class, Race, and Family Life*. Berkeley: University of California Press.

Leathwood, C., & Read, B. (2012). Final Report: Assessing the Impact of Developments in Research Policy for Research on Higher Education: An Exploratory Study. Society for Research in Higher Education. Retrieved December 31, 2019, from https://srhe.ac.uk/downloads/Leathwood_Read_Final_Report_16_July_2012.pdf.

Le-May Sheffield, S. (2004). *Women and Science: Social Impact and Interaction*. New Brunswick, NJ: Rutgers University Press.

Macdonald, A. (2014). *Not for People Like Me? Under-Represented Groups in Science, Technology and Engineering*. South East Physics Network: WISE. Retrieved December 31, 2019, from https://www.wisecampaign.org.uk/wp-content/uploads/2018/06/not_for_people_like_me-full-report.pdf.

Mandler, P. (2017). Educating the Nation: IV. Subject Choice. *Transactions of the Royal Historical Society, 27*, 1–27.

Manfredi, S., & Holliday, M. (2004). *Work-Life Balance: An Audit of Staff Experience at Oxford Brookes University*. Oxford: Centre for Diversity Policy Research.

Mare, R. D. (2011). A Multigenerational View of Inequality. *Demography, 48*(1), 1–23.

Marginson, S., Tytler, R., Freeman, B., & Roberts, K. (2013). STEM: Country Comparisons: International Comparisons of Science, Technology, Engineering and Mathematics (STEM) Education. Final Report. Australian Council of Learned Academies, Melbourne, Victoria. Retrieved June 3, 2018, from http://dro.deakin.edu.au/view/DU:30059041.

Martin, J. (2016). Education Policy. In A. Crines & K. Hickson (Eds.), *Harold Wilson: The Unprincipled Prime Minister – Reappraising Harold Wilson* (pp. 131–148). London: Biteback Publishing.

Mavin, S., & Bryans, P. (2002). Academic Women in the UK: Mainstreaming Our Experiences and Networking for Action. *Gender and Education, 14*(3), 235–250.

McIntosh, S. (2013). *Hollowing Out and the Future of the Labour Market*. BIS Research Paper Number 134. Department for Business, Innovation and Skills (DBIS). Retrieved September 18, 2016, from https://assets.publishing.service.gov.uk/government/uploads/system/uploads/attachment_data/file/250206/bis-13-1213-hollowing-out-and-future-of-the-labour-market.pdf.

McKie, R. (2019, August 31). Fears for Britain's Standing in World of Science as Students Shun Chemistry Degrees. *The Guardian*. Retrieved November 25, 2019, from https://www.theguardian.com/education/2019/aug/31/students-shun-chemistry-degrees-university-applications-fall.

McNeil, C. (2012). *The Politics of Disadvantage: New Labour, Social Exclusion and Post-crash Britain*. Institute for Public Policy Research (IPPR): Lankelly Chase Foundation. Retrieved September 19, 2016, from https://lankellychase.org.uk/wp-content/uploads/2016/05/IPPR_Report-Politics-of-Disadvantage-20121213.pdf.

Mills, C. (2008). Reproduction and Transformation of Inequalities in Schooling: The Transformative Potential of the Theoretical Constructs of Bourdieu. *British Journal of Sociology of Education, 29*(1), 79–89.

Morley, L., & Walsh, V. (1995). *Feminist Academics*. London: Taylor and Francis.

Morse, A. (2018). *Delivering STEM (Science, Technology, Engineering and Mathematics) Skills for the Economy*. Report by the Comptroller and Auditor General National Audit Office, Department for Business, Energy & Industrial Strategy, and Department for Education.

Morton, P., & Collins, J. (2010). Widening Horizons in Science, Technology, Engineering and Maths (STEM) Work Experience. Retrieved January 19,

2019, from https://warwick.ac.uk/fac/soc/ier/ngrf/stem/digging/equality/mortonandcolins_wideninghorizonsinstemworkexperience.pdf.
Munir, F., Mason, C., McDermott, H., Morris, J., Bagilhole, B., & Nevill, M. (2014). Evaluating the Effectiveness and Impact of the Athena SWAN Charter, Executive Summary, Equality Challenge Unit, Athena SWAN, Loughborough University. Retrieved March 3, 2019, from https://www.ecu.ac.uk/wp-content/uploads/external/evaluating-the-effectiveness-and-impact-of-the-athena-swan-charter.pdf.
National Union of Students (NUS). (2014). *Education Beyond the Straight and Narrow: LGBT Students' Experience in Higher Education*. London: NUS.
O'Hara, M. (2014). *Austerity Bites: A Journey to the Sharp End of Cuts in the UK*. Bristol: Polity Press.
Olsen, W., Gash, V., Heuvelman, H., & Walthery, P. (2014). The Gender Pay Gap in the UK Labour Market. In G. Razzu (Ed.), *Gender Inequality in the Labour Market in the UK*. Oxford: Oxford University Press.
Organisation for Economic Co-operation and Development (OECD). (2019). *Action Needed to Tackle Stalled Social Mobility*. OECD Home. Retrieved January 9, 2020, from https://www.oecd.org/newsroom/action-needed-to-tackle-stalled-social-mobility.htm.
Pakulski, J., & Waters, M. (1996). *The Death of Class*. London: Sage Publications.
Panel on Fair Access to the Professions (PFAP). (2009). *Unleashing Aspiration: The Final Report of the Panel on Fair Access to the Professions*. London: Cabinet Office.
Payne, G., & Abbott, P. (Eds.). (1990). *The Social Mobility of Women: Beyond Male Mobility Models*. Basingstoke: The Falmer Press.
Payne, G., & Roberts, J. (2002). Opening and Closing the Gates: Recent Developments in Male Social Mobility in Britain. *Sociological Research Online, 6*, 4, unpaged.
Pearce, R. (2017). Certifying Equality: Critical reflections on Athena Swan and Equality Accreditation, Centre for the Study of Women & Gender, University of Warwick. Retrieved June 3, 2018, from https://warwick.ac.uk/fac/soc/sociology/research/centres/gender/calendar/certifying_equality_a/certifying_equality_-_critical_reflection_on_athena_swan.pdf.
Pedley, R. (1956). *Comprehensive Education: A New Approach*. London: Victor Gollancz Ltd.
Perkin, H. (1989). *The Rise of Professional Society*. London: Routledge.
Pfeffer, F. (2014). Multigenerational Approaches to Social Mobility. A Multifaceted Research Agenda. *Research in Social Stratification and Mobility, 35*, 1–12.
Pickett, K., & Wilkinson, R. (2012, May 15). Sorry Nick Clegg – Social Mobility and Austerity Just Don't Mix. *The Guardian*. Retrieved July 18, 2018, from https://www.theguardian.com/commentisfree/2012/may/15/nick-clegg-social-mobility-austerity.

Powell, A. (2019). Unemployment by Ethnic Background. House of Commons Library. Briefing Paper No. 6385, 22nd May. Retrieved February 9, 2020, from https://researchbriefings.parliament.uk/ResearchBriefing/Summary/SN06385.

Powell, G., & Mainiero, L. (1992). Cross-Currents in the River of Time: Conceptualizing the Complexities of Women's Careers. *Journal of Management, 18*(2), 215–237.

Power, S., & Gewirtz, S. (*2001*). Reading Education Action Zones. *Journal of Education Policy, 16*(1), 39–51.

Rainsford, E., Rawlings, L., & Mistry, L. (2018). *The Role of Family in Social Mobility: Impact of Family Resources on Young People's Progression.* Youth Employment UK. Retrieved July 7, 2019, from https://www.youthemployment.org.uk/dev/wp-content/uploads/2018/03/family-Social-Mobility-Final3.pdf.

Reay, D. (2006). The Zombie Stalking English Schools: Social Class and Educational Inequality. *British Journal of Educational Studies, 54*(3), 288–307.

Reay, D. (2017). *Miseducation: Inequality, Education and the Working Classes.* Bristol: Policy Press.

Reay, D., Crozier, G., & Clayton, J. (2009). Strangers in Paradise: Working Class Students in Elite Universities. *Sociology, 43*(6), 1103–1121.

Roberts, G. (2002). *Set for Success: The Supply of People with Science, Technology, Engineering and Mathematics Skills.* Report of Sir Gareth Roberts' Review, HM Treasury. Retrieved August 12, 2016, from https://web.archive.org/web/20070205114712/http://www.hmtreasury.gov.uk/media/643/FB/ACF11FD.pdf.

Ryan, F. (2018, April 5). A Year of Disability Cuts Has Done Nothing But Starve the Sick. *The Guardian.* Retrieved November 19, 2019, from https://www.theguardian.com/commentisfree/2018/apr/05/year-disability-cuts-starving-sick-employment.

Sainsbury, D. (2007). *The Race to the Top: A Review of Government's Science and Innovation Policies.* HM Treasury: HMSO.

Sandberg, S. (2013). *Lean In: Women, Work, and the Will to Lead.* New York: Alfred A. Knopf.

Sandhu, S. (2019, November 28). Boris Johnson's Novel Seventy-Two Virgins Criticized for Racist and Sexist Language, Inews. Retrieved December 6, 2019, from https://inews.co.uk/news/politics/boris-johnson-novel-seventy-two-virgins-racist-sexist-language-1327014.

Santos, G., & Dang Van Phu, S. (2019). Gender and Academic Rank in the UK. *Sustainability, 11*(11), 1–46.

Savage, M., Devine, F., Cunningham, N., Taylor, M., Li, Y., Hjellbrekke, J., et al. (2013). A New Model of Social Class? Findings from the BBC's Great British Class Survey Experiment. *Sociology, 47*, 219–250.

Sellgren, K. (2017, January 26). *Social Mobility: Class Pay Gap Found in UK Professions*. BBC News, Family & Education. Retrieved June 17, 2018, from www.bbc.co.uk/news/education-38744122.

Simon, B. (1991). *Education and the Social Order 1940–1990*. London: Lawrence & Wishart.

Skeggs, B. (2002). Techniques for Telling the Reflexive Self. In T. May (Ed.), *Qualitative Research in Action* (pp. 349–375). London: Sage.

Skelton, C., & Francis, B. (2009). *Feminism and the Schooling Scandal*. London: Routledge.

Smail, B., Whyte, J., & Kelly, A. (1982). Girls into Science and Technology: The First Two Years. *School Science Review, 63*, 620–629.

Smith, E. (2010). Do We Need More Scientists? A Long-Term View of Patterns of Participation in UK Undergraduate Science Programmes. *Cambridge Journal of Education, 40*(3), 281–298.

Smith, E. (2011). Women into Science and Engineering? Gendered Participation in Higher Education STEM Subjects. *British Educational Research Journal, 37*(6), 993–1014.

Smith, M. (2014). *The Impact of Austerity on Schools and Children's Education and Well-Being*. The Encyclopaedia of Informal Education. Retrieved July 3, 2018, from http://infed.org/mobi/the-impact-of-austerity-on-schools-and-childrens-education-and-well-being/.

Smith, E., & White, P. (2011). Who Is Studying Science? The Impact of Widening Participation Policies on the Social Composition of UK Undergraduate Science Programmes. *Journal of Education Policy, 26*(5), 677–699.

Smith, E., & White, P. (2017). A 'Great Way to Get On'? The Early Career Destinations of Science, Technology, Engineering and Mathematics Graduates. *Research Papers in Education, 32*(2), 231–253.

Smith, E., & White, P. (2018a). *The Employment Trajectories of Science Technology Engineering and Mathematics Graduates, Final Report*. Leicester: University of Leicester/University of Warwick.

Smith, E., & White, P. (2018b). Where Do All the STEM Graduates Go? Higher Education, the Labour Market and Career Trajectories in the UK. *Journal of Science Education and Technology, 28*, 26–40.

Social Mobility and Child Poverty Commission (SMCPC). (2014). State of the Nation 2014: Social Mobility and Child Poverty in Great Britain, Report Summary. Retrieved July 16, 2017, from https://assets.publishing.service.gov.uk/government/uploads/system/uploads/attachment_data/file/367461/State_of_the_Nation_-_summary_document.pdf.

Social Mobility Commission (SMC). (2016). State of the Nation 2016: Social Mobility in Great Britain. Retrieved April 11, 2018, from www.assets.publishing.service.gov.uk/government/uploads/system/uploads/attachment_data/file/569412/Social_Mobility_2016_Summary_final.pdf.

Social Mobility Commission (SMC). (2019). State of the Nation 2018–19: Social Mobility in Great Britain. Retrieved December 12, 2019, from https://assets.publishing.service.gov.uk/government/uploads/system/uploads/attachment_data/file/798687/SMC_State_of_Nation_2018-19_Summary.pdf.

Solanke, I. (2011). Infusing the Silos in the Equality Act 2010 with Synergy. *Industrial Law Journal, 40*(4), 336–358.

Strelitz, J. (2004). Tackling Disadvantage: Education. In D. Darton & J. Strelitz (Eds.), *Tackling UK Poverty and Disadvantage in the Twenty-First Century An Exploration of the Issues* (pp. 57–75). London: Joseph Rowntree Foundation.

Sutton Trust. (2009). *Introduction in Social Mobility and Education: Academic Papers Presented at a High Level Summit Sponsored by the Carnegie Corporation of New York and the Sutton Trust, 2008.* Putney: Sutton Trust. Retrieved June 6, 2018, from https://www.suttontrust.com/research-paper/social-mobility-education/.

Swartz, B. (1997). *Culture and Power: The Sociology of Pierre Bourdieu.* Chicago, IL: University of Chicago Press.

Targeted Initiative on Science and Mathematics Education (TISME). (2013). What Influences Participation in Science and Mathematics? Retrieved August 28, 2015, from https://www.kcl.ac.uk/sspp/departments/education/research/aspires/TISME-briefing-paper-March-2013.pdf.

The Equality Trust. (2018). Socio-Economic Duty Unpaged. Retrieved December 6, 2019, from https://www.equalitytrust.org.uk/socio-economic-duty.

Thomas, E. (2001). *Widening Participation in Post-compulsory Education.* London: Continuum.

Thompson, P. (1997). Women, Men, and Transgenerational Family Influences in Social Mobility. In D. Bertaux & P. Thompson (Eds.), *Pathways to Social Class: A Qualitative Approach to Social Mobility* (pp. 32–61). Oxford: Clarendon Press.

Thompson, J., & Bekhradnia, B. (2009). Male and Female Participation and Progression in Higher Education. Full Report, Higher Education Policy Institute (HEPI) No. 41. Retrieved July 14, 2015, from http://www.hepi.ac.uk/2009/06/05/male-and-female-participation-and-progression-in-higher-education/.

Toffoletti, K., & Starr, K. (2016). Women Academics and Work–Life Balance: Gendered Discourses of Work and Care. *Gender, Work & Organisation.* https://doi.org/10.1111/gwao.12133.

Toogood, J. (2019). Civil Service: Inclusion in the Fast Stream, Blog. Retrieved January 6, 2020, from https://civilservice.blog.gov.uk/2019/09/27/inclusion-in-the-fast-stream/.

Tripney, J., Newman, M., Bangpan, M., Niza, C., MacKintosh, M., & Sinclair, J. (2010). *Factors Influencing Young People (Aged 14–19) in Education About STEM Subject Choices: A Systematic Review of the UK Literature.* Wellcome

Trust, EPPI Centre, Social Science Research Unit: Institute of Education, University of London.

Tyler, I. (2015). Classificatory Struggles: Class, Culture and Inequality in Neoliberal Times. *The Sociological Review, 63*, 493–511.

Universities UK (UUK). (2018). *Patterns and Trends in UK Higher Education*. London: Universities UK. Retrieved January 3, 2020 https://www.universitiesuk.ac.uk/facts-and-stats/data-and-analysis/Documents/patterns-and-trends-in-uk-higher-education-2018.pdf.

Vlaeminke, M. (1999). *The English Higher Grade Schools: A Lost Opportunity*. London: Woburn Press.

Walkerdine, V., Lucey, H., & Melody, J. (2001). *Growing Up Girl: Psychosocial Explorations of Gender and Class*. Basingstoke: Palgrave.

Wallace, M. (1993). Discourse of Derision: The Role of the Mass Media Within the Education Policy Process. *Journal of Education Policy, 8*(4), 321–337.

Waller, R., Harrison, N., & Last, K. (2015). Building a Culture of Participation: Interviews with the Former Directors of the National Aimhigher Programme. Retrieved June 3, 2019, from https://uwe-repository.worktribe.com/output/829528/building-a-culture-of-participation-interviews-with-the-former-directors-of-the-national-aimhigher-programme.

Watts, R. (2007). *Women in Science: A Social and Cultural History*. Abingdon: Routledge.

Weale, S., & Larsson, L. (2019, August 16). 'Students Don't See the Value': Why A-Level English Is in Decline. *The Guardian*. Retrieved December 3, 2020, from https://www.theguardian.com/education/2019/aug/16/students-dont-see-value-a-level-english-decline.

Weiner, G. (1985). *Just a Bunch of Girls: Feminist Approaches to Schooling*. Buckingham: Open University Press.

Wells, N., & Topping, A. (2010, August 18). Nick Clegg Attempts to Move Agenda to Social Mobility. *The Guardian*. Retrieved December 3, 2020, from https://www.theguardian.com/politics/2010/aug/18/nick-clegg-social-mobility.

Whitty, G. (2001). Education, Social Class and Social Exclusion. *Journal of Education Policy, 16*(4), 287–295.

Wilkins, S., Shams, F., & Huisman, J. (2013). The Decision-Making and Changing Behavioural Dynamics of Potential Higher Education Students: The Impacts of Increasing Tuition Fees in England. *Education Studies, 39*(2), 125–141.

Wilkinson, R., & Pickett, K. (2009). *The Spirit Level: Why Equality Is Better for Everyone*. Harmondsworth: Penguin Books.

Williams, J. (2011). STEM Education: Proceed with Caution. *Design and Technology Education, 16*(1), 26–35.

Wilson, H. (1963). *Labour's Plan for Science*. Reprint of Speech by Rt. Hon. Harold Wilson, M.P., Leader of the Labour Party, at the Annual Conference,

Scarborough, Tuesday October 1. Retrieved June 9, 2019, from http://nottspolitics.org/wp-content/uploads/2013/06/Labours-Plan-for-science.pdf.
Women into Science Technology, Engineering and Mathematics (WISE). (2014). *The Talent Pipeline from Classroom to Boardroom*. UK Statistics 2014. WISE: Campaign for Gender Balance in Science, Technology & Engineering. Retrieved December 3, 2019, from https://wisecampaign.scdn3.secure.raxcdn.com/wp-content/uploads/2018/04/WISE_UK_Statistics_2014.pdf.
Woodin, T., McCulloch, G., & Cowan, S. (2013). *Secondary Education and the Raising of the School-Leaving Age: Coming of Age?* London: Palgrave Macmillan.

Index[1]

A
Abbot, P., 25
Adonis, A., 43
AdvanceHE, 2, 46, 47, 100
  leading partner in advancing equality, 46
Advanced level results, *see* Examination results
American Association of University Women, 36
Andrews, J., 3, 26, 27
Apprenticeship levy, 29
  low-skill jobs rebadged, 29
Archer, L., 3, 8–10, 39–42, 86, 107–109, 127, 135–137, 143
Ashenden, A., 44
Aspirations, *see* Student aspirations
Aspires project, 39, 127
Athena Swan Charter (ASC)
  agreed principles, 46
  award levels, 47, 90, 105, 107, 116
  business and law, 47
  concerned with people's representation, 47
  extended to cover arts, 47
  humanities, 47
  lacks systemic control, 46
  privileges whiteness and middle class, 47
  progression of students into academia, 47
  role of ASC in case study department, 13, 90
  social sciences, 47
  2013 and 2018 reviews, 48
Atkinson, A., 26, 76
Austerity
  creates harsh social climate, 28
  social mobility and austerity don't mix, 29
Avendano, L., 31, 38

B
Background circumstances, 19, 26, 134
  influence on young people, 26
Bagilhole, B., 91, 123

[1] Note: Page numbers followed by 'n' refer to notes.

K. Hoskins, B. Barker, *STEM, Social Mobility and Equality*,
https://doi.org/10.1007/978-3-030-49216-8

Ball, S., 9, 127, 136
process of acquiring dispositions, 9
Ballarino, G., 137
Barker, B., 3, 4, 25, 26, 28, 29, 39, 71, 76, 135, 140
Barnett, C., 20
BBC Class Survey, 26
Bell, D., 32
Belmi, P., 109
Benn, C., 21
Bernardi, F., 137
Bertaux, D., 2, 5, 6, 67
Bhopal, K., 2, 46–48
Bilge, S., 43
Black and Minority Ethnic (BME)
less likely to be in senior positions, 48
promotion barriers, 48
women less likely to be professors, 48
women most disadvantaged, 48
Blair, T., 22, 23
the 'knowledge economy, 22
revolution in technology speech, 22
social mobility a great force for social equality, 23
target for higher education growth, 22
Blanden, J., 11, 23, 27
intergenerational income mobility low and falling, 11, 23
Bloome, D., 42, 43
Boliver, V., 6
Bourdieu, P., 8–10, 25, 26, 40, 41, 72, 135, 137
conceptual framework, 8
field, 9
habitus and dispositions, 8
Breiner, J., 30
Bridge Group
Civil Service, 142, 143
Fast Stream, 142
internship programme, 142
promotes socio-economic diversity, 142
removal of candidate screening criteria, 142
British Cohort Study (BCS), 6, 11, 11n1
British Household Panel Survey (BHPS), 11n1
British Social Attitudes Survey, 63
Broecke, S., 33
Brown, D., 27
Brown, G.
worries about exclusion and privilege, 23
Brown, P., 28, 29, 39, 140
Buckingham, J., 48, 94, 96
Bukodi, E., 27, 135–137
Bunting, M., 27
Burr, V., 12, 97, 102, 109
Butler, P., 63
Byrne, A., 25

C

Canetto, S., 122
Capital, 130
acquisition and role of science capital in forming identities and aspirations, 13, 62
impact on family and individual employment trajectories, 4
role in production of social advantage and disadvantage, 10
science capital definition, 10, 66
social and cultural resources, 9
transmission of economic, 9, 78
Caring duties, 93, 120, 138
Case study
staff participants; Audrey, 92, 104–106, 123, 125–128; Ellen, 91, 103, 126; Hayley, 92, 99,

100, 103, 105, 117, 119, 124, 127; Linda, 98–102, 106, 115, 128; Noah, 93–95, 98, 106, 109; Rachel, 95, 98, 106, 116, 119, 124, 126, 138; Robin, 91, 93–97, 101–103, 106, 117–119, 121–125, 128, 136, 138, 139, 141; Susie, 92, 96, 99, 102, 105, 116, 120, 121, 123, 138
student participants, 7; Ainsley, 63–65, 67, 69, 70, 75, 77–79, 81, 83–85, 120; Anna, 63–65, 67, 69, 70, 72–74, 77, 79, 82, 84–86; Bella, 63, 65, 69, 72, 74, 79–81, 83–85; Carrie, 63, 66, 68–70, 75, 77–80, 82, 83, 85; Daniella, 63–65, 68, 69, 72, 73, 77, 79–81, 83, 85, 86; David, 63–65, 68, 69, 71, 72, 75, 77, 80, 83; Gabriella, 63, 64, 66, 69–71, 76–80, 82–85; Gail, 63, 66, 69, 70, 72–74, 77, 78, 80, 82–84, 86; George, 63, 65, 67, 70–72, 78, 80, 82, 83, 85, 120; Joel, 63, 65, 68, 70, 76, 78, 80, 82, 84, 85; Kyle, 63–65, 70, 71, 75, 77, 79, 80, 82
Case study design
Ethics, 8
Interviews, 2
propositions to explain data, 48
sample selection, 7
Cassidy, R., 38
Cavendish laboratories, 20
Centre for Education in STEM education (California State University), 38
establishes WISE programme in the US, 38
Chan, T. W., 6
Chemistry students, 2, 4, 6–9, 13, 40, 62, 84, 90, 94, 100, 102, 114, 115, 118, 120, 123, 136, 138
family occupations and class identities, 148–149
interviews, 7, 90
sample, 7
Chilisa, B., 12, 138
Chitty, C., 21
Class, *see* Social class
Collini, S., 141
Collins, R., 27, 32, 43
Commission for Racial Equality, 45
Comprehensive schools
campaign for, 21
circular 10/65, 21
ending selection at age 11, 21
Cook, C., 3, 24, 27
Cornwell, J., 20
Cotterill, P., 121
Crenna-Jennings, W., 26, 27
Crozier, G., 9, 71, 72, 76, 90n1
Cultural, *see* Capital

D
Dang Van Phu, S., 121
David, M., 63–65, 68, 69, 71, 72, 75, 77, 80, 82, 83, 121
Department for Children
role in improving STEM delivery, 33
Department for Children Schools and Families (DCSF), 32
Dex, S., 25
Disability Rights UK
employment gap between disabled and non-disabled, 43
handbook for campaigners, 43
Disadvantage
impact on equal opportunities and social justice, 26

Disadvantage (*cont.*)
'legacies of disadvantage,' 42
linked with poor educational outcomes, 26
mobility for talented few, 42
persistence of large scale inequality and disadvantage, 26
reduced access and opportunity, 28
Disadvantaged students
austerity reduces opportunity, 22
gaps in HE entry rates between advantaged and disadvantaged, 35
given inadequate access to STEM content in US, 31
lack of diversity in US college majors, 31
less likely to progress in education and careers, 26
lower progression rates from sixth form to most selective universities, 35
more likely to be in further education than sixth forms, 28
more likely to be in low-level service jobs, 28
Dispositions, *see* Student dispositions
Dorling, D., 25
Dryden, S., 43, 44
Dyhouse, C., 25

E

Ebdon, Les, 139
Economic capital, *see* Capital
Education
impact on students' ability to transcend disadvantaged circumstances, 4
policy changes, 94–103
outcomes follow patterns of wealth and family advantage, 29
Education Maintenance Allowance, 140
Education reform, 23, 26, 28
failing to compensate for class structure or life circumstances, 28
Eleven+ examination
children excluded from grammar schools and qualifications, 20
Equality, 2, 12, 13, 23, 24, 35, 42–48, 78, 86, 90–109, 113–130, 134, 135, 139–141
Equality Act (2010)
adopts unitary perspective, 45
enforced by single commission, 45
expands positive duties on public authorities to advance equality, 45
implements EU directives, 45
protected characteristics, 43, 45, 118, 130
replaces previous legislation, 45
retains logic of separate discrimination claims, 45
scope for transformative equality, 141
social class not a protected characteristic, 48
socio-economic duty not implemented, 45, 129, 130, 141
Equality and Human Rights Commission (EHRC), 44, 45, 141, 142
Equality Challenge Unit (ECU) (now AdvanceHE)
commissions review of charter impact, 47
leading partner in advancing equality in further and higher education, 46
Equal opportunities

constrained by class structure, 26
to enable people to overcome disadvantage, 23
policy priority for Wilson and Blair to reduce wasted talent, 22
seriously unequal opportunities continue, 37, 137
Equal Opportunities Commission, 38, 45
Equal rights, 43, 45, 48, 141–143
depend on favourable circumstances, 43
not easily achieved, 43
Erikson, R., 25
Erola, J., 6
Ethnicity, *see* Race and ethnicity
Examination results
reflect distribution of wealth, 26
relationship with economic disadvantage, 13

F

Family background, 4, 6, 10, 26, 48, 62–69, 71, 126, 127, 129, 140
influences on education and vocational choice, 4
Family capital, *see* Capital
Family habitus, *see* Habitus
Family influences
impact of family context on science attitudes, 40
impact on career and educational trajectories, 25
inter-action of habitus and dispositions with science capital, 40
role in shaping aspirations and goals, 4
Family networks
influence on choice of academic and vocational pathways, 5, 62, 86
role in holding and transmitting capital, 5
Father-son income differences, *see* Male income differences
Felix Holt, 4
Finegold, P., 39
Flexible working, 90, 96–100, 107, 115, 117, 138
Fowler, S., 36, 38
Francis, B., 12, 37
Free school meals (FSM), 26, 35, 105, 129
eligibility, 129
impact on examination results, 129
Friedman, S., 3, 129, 135, 136, 140, 143

G

Gender
chemistry recruitment gender neutral, 36
female careers and options limited by gendered constraints, 37
impact on access, 86, 114
influence on subject choice and aspirations, 12
opportunity and progress, 7
shapes life and life chances, 27
social construction of, 12, 97, 137
Genealogical case study, 13, 62
eleven families through three generations, 13, 62
Genealogical methods
habitus and dispositions in building aspirations and science capital, 61
investigating family networks, 61
role of family networks, 5

General Certificate of Secondary Education (GCSE), 21, 71–76, 85
introduced, 21
results (*see* Examination results)
Generalization, 2
Gewirtz, S., 23, 29
Girls into Science and Technology (GIST), 38
aims to increase female participation, 38
Gladwell, M., 86, 140
Goldin, C., 3
Goldthorpe, J., 11, 25, 27, 135–137
Gorard, S., 11
Goulden, C., 42
Gove, M., 24, 39
education and social mobility, 24
emphasis on importance of equality, 24
school reforms to improve outcomes, 24
Government policies, 2, 12, 39, 48, 103, 136, 137
resistances to equal opportunities and social mobility, 2, 39
Grandparents, 4, 6, 7, 64, 65, 68, 70, 84
role in mobility, 6
Greenfield, S., 37, 83, 93, 135
Greening, J., 24
plan to improve social mobility through education, 24

H

Habitus, 40, 42, 83, 85, 86, 127, 135–137
definition, 11
differences in habitus produce uneven class and race-based patterns in children's science aspirations, 40
influence on choice of academic and vocational pathways, 5, 62
relays and reproduces social class and inequality, 42, 127
Haldane, J., 20
Hamed, J., 33
Hazell, W., 141
Hepple, B., 45, 141
Higher education (HE)
entry stratified by class, 27
expansion, 3, 4, 21, 23, 26, 27, 33, 35
expansion working more favourably for better off students, 23
growth in science and science-related subjects, 33
increased male/female participation, 33
Robbins Report, 21
US funding to increase participation in STEM, 31
Higher grade schools, 20
Holliday, M., 116
Holman, J., 39
Hoskins, K., 3, 4, 25, 26, 28, 29, 39, 71, 76, 95, 98, 121, 128, 135
Human capital theory, 3, 23
influence on policy, 23

I

Identities, *see* Student identities
Inclusion, *see* Social inclusion
Individual agency
constrained by family and social and economic structures, 26
need to pay more attention to families and households, 26
personal agency possible without knowledge and resources at home, 41
Individualism, 4

role in government social mobility policy, 4
Inequality
'discourse of derision' to include class, gender, race, ethnic minorities and LGBT, 42
flawed policies to tackle, 28
gaps in HE entry rates between advantaged and disadvantaged, 35
impact and effects, 42
unequal access and progression in higher-level salaried positions, 135
unequal mobility chances persistent, 135
unequal structures condition perceptions of what is possible, 42, 138
widening access has minimal impact on unequal outcomes for less advantaged, 35
Institute of Careers Guidance Award, 32
Intergenerational mobility, 6
International competition
advanced nations aim to move to high value goods to compete with low-wage economies, 30
education reforms to equip young people to compete, 23
growth and impact, UK drive to improve competitiveness, 22
policies adopted globally to increase, 30
Interviews, *see* Case study design; Chemistry students

J
Joannou, M., 43
Jones, O., 24, 42, 43

K
Kahn, O., 43
Kaiser Wilhelm Institutes, 20
Keck, T., 43, 44
Kinship ties, *see* Family networks
Kuenzi, J., 30, 31, 33

L
Labour's Plan for Science, 3, 20
Lambert, P., 5, 11, 25
Lammy, David, 139
Lareau, A., 40, 76
Larsson, L., 34, 104
Laurison, D., 3, 129, 135, 137, 140, 143
Leathwood, C., 116
Lesbian, Gay, Bi-Sexual and Transgender (LGBT)
campaigns, 44
government action plan, 44
homelessness, 44
homophobic behaviour, 44
persistent prejudice, 44
Life chances, 4, 26–29, 86
not helped by widening participation, 4
Long-hours culture, 117, 138
Luckmann, T., 12

M
Machin, S., 23, 27
intergenerational income mobility low and falling, 23
Mainiero, L., 83
Major, J.
Further and Higher Education Act (1992), 21
Male income differences, 11
usefulness as intergenerational mobility measure, 6
Mandler, P., 4, 135

Manfredi, S., 116
Mapping Review, 32
tracking STEM initiatives, 32
Mare, R., 6, 86
Marginson, S., 30, 37
Marsden chemistry department
ASC awards, 107
barriers to equality, 134
curriculum design, 134
equality and diversity, 134
high quality teaching, 134
inclusion and diversity, 134
inclusive culture, 139, 142
'leaky' pipeline effect, 135
opportunities for career progression, 134
Marsden University, 1, 7, 14, 133
Martin, J., 21
McIntosh, S., 28
McNeil, C., 22, 23, 26
Meritocratic society, 3, 29
Middle class, *see* Social class
Millennium Cohort Study (MCS), 11n1
Mills, C., 10, 27, 76
Moisio, P., 6
Morley, L., 116
Morse, A., 31
Morton, P., 32
Multi-generational effects, 6
Munir, F., 47

N

National Child Development Study (NCDS), 11, 11n1
National Curriculum, 36
National Institutes of Health (US) (NIH), 33
receiving funds to increase participation, 33
National Science Foundation (US) (NSF), 30, 33
receiving funds to increase participation, 33
National Science Learning Network, 32
National Union of Students (NUS) (2014), 120
New Labour
assumption higher-level skills needed, 30
emphasis on targets and delivery, 24
fails to increase flow of socially mobile and skilled workers, 29

O

Office of Science and Innovation (OSI), 32
O'Hara, M., 28, 39, 44, 140
Olsen, W., 3, 25
Organisational culture
works against women, 37

P

Pakulski, J., 27
Parental class, 27
strong influence on academic attainment, 27
Parsons, V., 44
Pay gap for women, 3
Payne, C., 25
Payne, G., 25, 26
Pearce, R., 134, 139, 140
Pedley, R., 21
Peers, 26
Performance gap, 137
between advantaged and disadvantaged, 137
Pfeffer, F., 6
Pickett, K., 24, 29, 39, 42, 140

POLAR measure, 141n1
Policy agenda, 22
  drive to increase STEM recruits, 22
Poverty, *see* Disadvantage
Powell, G., 83
Protected characteristics, *see* Equality Act (2010)
Purvis, J., 43

Q

Quantitative surveys, 5
  limitations on usefulness, 24–26

R

Race, *see* Race and ethnicity
Race and ethnicity
  African-Americans, 31
  BAME unemployment levels, 44
  greater likelihood of experiencing childhood deprivation and murder, 44
  Hispanics and native Americans a small minority of STEM students in US, 31
  impact on opportunity, 44
  incidence of child deprivation amongst ethnic minorities, 44
  racist stereotypes, 42
Race Equality Charter (REC)
  encourages cultural and systemic changes, 46
  five guiding principles, 47
  specific charter for race equality, 46
Raising of the school leaving age (ROSLA), 21
Read, B., 116
Reay, D., 27, 106, 109, 136
Robbins Report, *see* Higher education (HE)
Roberts, G., 31
Roberts, J., 25, 26
Royal Society of Chemistry, 35
Russell Group, 1, 5, 61, 67n2, 72, 77, 107, 108, 115, 129, 136, 141
Ryan, F., 44

S

Sainsbury, D., 30
Sandhu, S., 42
Santos, G., 121
Savage, M., 39
School reforms, 4, 33
  ability to increase social mobility, 24
Science and Art Department examinations, 20
Science capital, *see* Capital
Science education
  extending and improving provision, 21
  not equipped for the modern age, 21
Science, Technology, Engineering and Mathematics (STEM)
  acronym introduced, 30
  drive to increase numbers, 22
  drop in number of chemistry applicants, 135
  fall in numbers, 31
  female students outnumbering males in A-level science entries, 34
  gender bias in subjects, 12
  initiatives in UK to increase participation by under-represented groups, 33
  labour market conditions not consistent with STEM shortage, 34
  overall growth in male/female numbers choosing STEM subjects, 33

Science, Technology, Engineering and Mathematics (STEM) (*cont.*)
privileged position in UK HE policy, 32
Roberts Review identifies downward trend in numbers taking STEM qualifications, 31
shortage of STEM graduates willing to work in highly skilled STEM roles, 34
shortages in Australia and UK, 31, 33, 34
STEM initiatives have limited impact on long-term A-level recruitment, 35
strategic role of science and technology, 20
top-funding priority in the US, 30
UK school reforms to raise standards and STEM participation, 33
women and ethnic minorities under-represented in science and engineering, 31, 33
worries about shortage of STEM skills in UK, US, Australia and EU, 30, 31, 33, 34
Scientific Revolution, white heat of, 20–21
Scientific Women's Academic Network (SWAN), 46
Section 28, 1988 Local Government Act, 44
preventing promotion of homosexuality and 'pretended family relationships,' 44
Sellgren, K., 27
Seventy-Two Virgins, 42
racist and sexist stereotypes, 42
Sheffield Hallam University, 32
Education Business Partnership, 32
Simon, B., 20, 21
Skeggs, B., 8
Skelton, C., 12
Smith, E., 4, 26, 27, 32–37, 48, 90, 91, 95, 103, 107, 126, 141
Smith, M., 28
Social capital, *see* Capital
Social class, 14, 21, 27, 28, 40–42, 63–66, 90n1, 101, 106–109, 115, 122, 126–130
background, 2, 7, 9, 13, 62, 94, 105, 107, 115, 126–129, 140
class inequalities load dice, 28
demonization of the working class, 42
extra resources and activities for middle class children, 27
impact on prospects for intergenerational social mobility, 9
intersections with gender, 9
middle class background facilitates potential science aspirations, 41
not included as a protective characteristic under the Equalities Act (2010), 2, 48, 115, 142
pay gap between classes, 27
rates of class mobility, 27
role in formation of aspirations to study chemistry, 5
shapes life chances, 27
socio-economic gap in university participation, 35
working and middle class life patterns different, 27
working class children miseducated, 27
Social inclusion, 2, 3, 8, 23, 26, 104, 107–109, 128, 134, 138, 139, 141
Social mobility
dominant policy theme influencing approach to equality issues, 23

evidence of poverty and inequality leads policies to improve mobility, 24
government expectations unrealistic, 4
'low and falling,' 11, 23
methodological issues, 11
mixed picture of mobility from studies, 25
policy not overcoming barriers, 29
problems with government policies, 2, 39
rates of class mobility, 27
research flawed, 24–26
role of housing as barrier, 42
Social Mobility Commission (SMC), 3, 24, 26–29, 42, 129, 135
assessment of progress towards fairness, 29
Social reproduction
class and inequality woven into young people's dispositions, 41
intergenerational mobility, 9
Socio-economic duty, 45, 129, 130, 141–143
South Park, 4
Special Educational Needs (SEND), 26
associated with poverty, 26
Staff participants
academic culture of overworking, 117–118, 121
academic role models, 101
action plan for ASC, 113–114
adequacy of support groups, 120
aim to diversify student and staff body, 90, 95
aim to include a greater number from minority and working class backgrounds, 101
alignment of existing culture with ASC, 94
areas for further action to enhance equal opportunities, 114
balance between selection and inclusion, 108
balancing act between work and home life, 98, 121
bridges for students from disadvantaged backgrounds, 104
certain groups lack dispositions to study chemistry, 126
challenge to recruit BAME students, 123, 139
challenging stereotypes, 120
charter framework, 114
chemistry course intensive and demanding, 108
Chinese students disrupt trend, 124
choosing post codes for outreach, 141
contribution to tackling unequal representation of women in science, 100
debt-averse attitudes stronger for lower-class students, 128
decision to apply for recognition, 94
department 'very white, middle class,' 106
difference in workload profiles of high-status and low-status professors, 116
dilemmas for part-time staff seeking promotion, 115
disappointments and dilemmas, 114
disproportionate caring duties, 93, 120, 138
diversifying chemistry, 118–129
dominant view – males will be scientists, 103
dominant work culture demands overwork, 117

Staff participants (*cont.*)
drop-off continues despite maternity leave for post-doctoral researchers, 96
ease of applying for part-time working and return to full time, 97
efforts to increase appointments from under-represented groups, 100
elite institutions lack social diversity, 106
equality and diversity policy, 78, 95
equality at Marsden, 90
equality policies, 13, 90
ethnicity, 114
ethos of taking extra step, 95
external examiners from varied backgrounds, 102
family friendly policy, 92
financial concerns, 127
five key priorities, 113
fixing the 'leaky' pipeline, 103–104
flexibility aids work-life balance, 99
flexible working initiative, 96–100, 115
focusing on representation, 101–103
gender, 90, 91, 93, 103, 114, 139
gendered caring commitments, 121
gender equity, 90, 93, 102
gender/ethnicity/social class intersections, 94
gender representation, 91, 139
genuine mix of guest speakers, 101
high drop-out rate after doctorate, 96, 103
inclusive and enabling environment, 91
inclusive working ethos, 93
increases in female academic staff, 114
inequality persists at Marsden, 129
initiatives to improve diversity, 101
intractable nature of social class, 126
lack of female role models, 103
lack of wider institutional support, 118
'leaky' pipeline persists, 104
LGBTQI+ support, 118, 119
limitations of the Equalities Act (2010), 115
limits of equality, 129
long-hours culture, 117, 138
loss of women from careers in chemistry, 138
low aspirations, 126
Marsden weak on ethnic minority background, 118
masculine hyper-individualised work cultures, 122
maternity leave policy for doctoral students, 96
mega star research professor part-timer, 117
minority ethnic students more likely to live at home, 105
natural male scientist, 109
need for changes in society and wider culture around female academics in chemistry, 121
need for further diversification, 118
need to implement socio-economic duty, 143
new policies to comply with ASC principles, 94
non-hierarchical working culture, 90
'not sure that's right for you,' 127
NUS Report, 120
obtaining an award easier than for traditional departments, 91
occupational class distribution unchanged, 35
on-campus nursery, 92

parental leave taken further than ASC, 96
partnership with London sixth form college, 105
patterns of recruitment, 107
PhD/post-doc barrier, 96, 103, 108, 121
policy changes and initiatives, 94–103, 115
pressures for women scientists, 123
promotion processes, 115–117
'racehorses and carthorses,' 116
raising aspirations, 105
rebalancing socially constructed gender responsibilities, 97
reducing stress and anxiety, 98
role of geography, 73
role of habitus, 124
role of habitus and dispositions, 5, 9, 85
role of social class background, 7
school origins of students, 129
science-specific expectations, 40
selection v. inclusion at Russell Group universities., 108
shared parental leave policy, 95–96
shortlisting to ensure fair representation, 102
slow progress towards a more socially inclusive chemistry department, 109
social class background, 94, 105, 126
social class representation, 105–107
some academically capable students struggling, 117
some parents discouraging, 143
staff appointment panels, 102
staff appointment practices, 100–101, 138
staff workload models, 116
stereotyping, 120
support for women's careers, 93
supportive family policies, 90
unconscious bias training, 100, 102
under-representation of staff and students from minority ethnic and disadvantaged backgrounds, 129
'university is not for me,' 126
widening participation policies, 90, 104–105
women learning self-promoting techniques, 123
women well-represented, 36, 114
'work doesn't go away,' 117
workshops, 139
worries about fair distribution of work across teams, 118
Starr, K., 116
Stereotypes, 42, 43, 119, 120, 137, 139
Strelitz, J., 43, 44
Student aspirations, 2–6, 10, 13, 14, 24, 32, 39–42, 48, 62, 64–72, 79, 82, 90, 105, 109, 115, 124–127, 130, 133, 135, 136, 138, 143
Student dispositions, 4, 5, 7, 10, 40–42, 61, 62, 66–68, 71, 72, 76, 83–86, 108, 126–129, 135–137
impact on academic and employment trajectories, 62
Student identities, 5, 12–14, 37, 43, 47, 48, 62, 71, 72, 76, 82, 84, 115, 127, 142
formed through families, 13
habitus and dispositions, 62
impact on choice of education and vocational pathways, 86
Student participants
ambitions revised pragmatically, 78
aspirations, 62

Student participants (*cont.*)
building on advantages, 86
building science capital, 61, 85
bursary to work on admissions, 80
careers symposium, 80
challenge of combining children and a career, 82
chemistry degrees, 65
choir and brass band, 80
choosing university, 77–79
class origins and upward mobility, 63–65
committed to science careers in higher education, 81, 84
committed to scientific careers, 84
concerted cultivation, 76
contrasting profiles of mothers and grandmothers, 64
department encouraging, 79
desire for high status and wealth disavowed, 69
diversity in culture, 67
emphasis on happiness and job satisfaction, 85
emphasis on personal happiness, 69
encouraging, 68
encouraging self-reliance and hard work, 85
families source of disposition, 68
families upwardly mobile, 67
family culture working conservatively, 71
family networks, 66, 86
final decisions postponed until after results, 84
financially secure backgrounds, 84
future plans, 82, 83, 85
genealogical perspective, 86
grandparents' occupations, 65
habitus and dispositions shape choices, 84
identities as bright and mathematical/scientific, 85
impact of course demands on other interests, 79
impact of staff, 46
individual agency constrained, 86
individual histories, 67
industry placements challenging, 79
influence of family models and networks, 85
intergenerational change, 66
intergenerational transmission consistent with Thompson families, 66
internships, 79
job in industry, 82
job satisfaction and desire to make a difference, 69
knowledge and capitals deployed to ensure success, 76
lecturers excellent, 79
M. Chem., 79
migrant experiences, 66
most common profile middle class, 64
motivation and aspirations, 68, 69
occupational influences, 65
open day, 68, 77, 85
other family influences, 67–68
outstanding academic ability, 67
partners and children, 82
PhD, 81
preference for Russell Group, 77
preparation for PhD, 80
research, 82
role of competition, 70
role of parents, 78
role of subject, 34, 72
sample profile consistent with TISME findings on source of positive attitudes and dispositions in science, 66

school and university enablers, 86
school education, 74
schools challenging for some, 74
strong family influences on lives and learning, 71
strong STEM influences, 65
summer projects, 79
supportive families, 68
university experience, 70, 79–81
values and strategies, 67
volunteering, 81
Suffragette movement
difficulty of moving beyond recognition and formal legal rights, 43
Sure Start, 140
Sveinsson, K., 43
Swartz, B., 84, 137

T
Targeted Initiative on Science and Maths Education (TISME)
family science capital predisposes students to STEM, 40
identifies attitudes and resistances, 66
maps young people's aspirations and engagement with science, 39
Ten-Year Science and Innovation Investment Framework (SIIF), 32
Thompson, P., 5, 6, 66, 67, 71, 136
Toffoletti, K., 116
Tomei, A., 107, 109
Toogood, J., 143
Topping, A., 11
Top professions
dominated by private school, 3
Tripney, J., 33
Tuition fees, 28, 128
Tyler, I., 42, 43

U
University tuition fees, *see* Tuition fees

V
Vlaeminke, M., 20

W
Walkerdine, V., 27, 37, 38
Waller, R., 106
Walsh, V., 116
Waters, M., 27
Watts, R., 20
Weale, S., 34, 104
Weiner, G., 12
Wells, N., 11
White, P., 26, 27, 32, 34, 36, 41, 44, 48, 90, 91
Widening participation
benefitting better off, 27
educational and vocational outcomes unchanged, 35
increased participation in higher education, 22
increased sixth form numbers, 21, 22, 29
increases in female participation in OECD countries, 33
Marsden policy, 90, 104–105
number of universities doubled, 21
patterns of participation amongst UK undergraduates stable by background and ethnic group, 34
proportion of disadvantaged students at university below target, 139
raising the school leaving age, 21
role of John Major and Tony Blair, 21

Widening participation (*cont.*)
world-wide investment to increase participation in science and technology, 30
Wider Horizons initiative
work experience to raise aspirations and widen access, 32
Wilkins, S., 128
Wilkinson, R., 24, 29, 39, 42, 140
Williams, J., 34
Wilson, H., 3, 20–22
'brain drain,' 21
comparisons with Russia, 21
demands more scientists, 21
Scarborough speech, 21
wasting talent, 21, 22
Women
careers hindered by traditional family roles and responsibilities, 93
chemistry recruitment gender neutral, 36
constrained by gendered hierarchies, 37
contribution to households and mobility, 64
employment, 5, 25
especially in senior posts, 37
female candidates equally interested and able in science, 36
female options and careers limited by gendered constraints, 37
female participation in US and UK STEM degrees and jobs, 38
female recruitment to physics and engineering small, 36
growth of female employment, 25
intersections of race, class, gender, sexuality, disability and age, 43
'leaky pipeline' effect in women's careers, 38
need to consider women in terms of their own social class and occupation, 25
pay gap between women and men, 37
role in raising children, 6
role in transmission of family influences, 6
status and earnings, 5
treatment in mobility research, 25
under-represented in science and engineering, 31
women under-represented in managerial and professional career, 37
Women into Science and Engineering (WISE)
despite initiatives gender gap in maths and physics at A level persists, 38
mission to engage women in science and engineering and improve recruitment, 38
Woodin, T., 21
Woodward, D., 121
Working class, *see* Social class
Workload, 98, 99, 108, 115–118, 138, 140

The manufacturer's authorised representative in the EU is Springer Nature Customer Service Centre GmbH, Europaplatz 3, 69115 Heidelberg, Germany. If you have any concerns regarding our products, please contact ProductSafety@springernature.com

Printed and bound by CPI Group (UK) Ltd, Croydon, CR0 4YY
30/09/2026
02225645-0001